AF868416

Mechanisms and Machine Science

Volume 130

Series Editor

Marco Ceccarelli, Department of Industrial Engineering, University of Rome Tor Vergata, Roma, Italy

Advisory Editors

Sunil K. Agrawal, Department of Mechanical Engineering, Columbia University, New York, NY, USA

Burkhard Corves, RWTH Aachen University, Aachen, Germany

Victor Glazunov, Mechanical Engineering Research Institute, Moscow, Russia

Alfonso Hernández, University of the Basque Country, Bilbao, Spain

Tian Huang, Tianjin University, Tianjin, China

Juan Carlos Jauregui Correa, Universidad Autonoma de Queretaro, Queretaro, Mexico

Yukio Takeda, Tokyo Institute of Technology, Tokyo, Japan

This book series establishes a well-defined forum for monographs, edited Books, and proceedings on mechanical engineering with particular emphasis on MMS (Mechanism and Machine Science). The final goal is the publication of research that shows the development of mechanical engineering and particularly MMS in all technical aspects, even in very recent assessments. Published works share an approach by which technical details and formulation are discussed, and discuss modern formalisms with the aim to circulate research and technical achievements for use in professional, research, academic, and teaching activities.

This technical approach is an essential characteristic of the series. By discussing technical details and formulations in terms of modern formalisms, the possibility is created not only to show technical developments but also to explain achievements for technical teaching and research activity today and for the future.

The book series is intended to collect technical views on developments of the broad field of MMS in a unique frame that can be seen in its totality as an Encyclopaedia of MMS but with the additional purpose of archiving and teaching MMS achievements. Therefore, the book series will be of use not only for researchers and teachers in Mechanical Engineering but also for professionals and students for their formation and future work.

The series is promoted under the auspices of International Federation for the Promotion of Mechanism and Machine Science (IFToMM).

Prospective authors and editors can contact Mr. Pierpaolo Riva (publishing editor, Springer) at: pierpaolo.riva@springer.com

Indexed by SCOPUS and Google Scholar.

Marco Ceccarelli · Loredana Santo ·
Marco Paoloni · Giacomo Cupertino
Editors

Design Advances in Aerospace Robotics

Proceedings of TORVEASTRO 2023

IFToMM

Springer

Editors
Marco Ceccarelli
Department of Industrial Engineering
University of Rome Tor Vergata
Rome, Italy

Loredana Santo
Department of Industrial Engineering
University of Rome Tor Vergata
Rome, Italy

Marco Paoloni
Robotics and Artificial Intelligence
Laboratory
ENEA
Rome, Italy

Giacomo Cupertino
Robotics and Artificial Intelligence
Laboratory
ENEA
Rome, Italy

ISSN 2211-0984 ISSN 2211-0992 (electronic)
Mechanisms and Machine Science
ISBN 978-3-031-28446-5 ISBN 978-3-031-28447-2 (eBook)
https://doi.org/10.1007/978-3-031-28447-2

© The Editor(s) (if applicable) and The Author(s), under exclusive license to Springer Nature Switzerland AG 2023
This work is subject to copyright. All rights are solely and exclusively licensed by the Publisher, whether the whole or part of the material is concerned, specifically the rights of translation, reprinting, reuse of illustrations, recitation, broadcasting, reproduction on microfilms or in any other physical way, and transmission or information storage and retrieval, electronic adaptation, computer software, or by similar or dissimilar methodology now known or hereafter developed.
The use of general descriptive names, registered names, trademarks, service marks, etc. in this publication does not imply, even in the absence of a specific statement, that such names are exempt from the relevant protective laws and regulations and therefore free for general use.
The publisher, the authors, and the editors are safe to assume that the advice and information in this book are believed to be true and accurate at the date of publication. Neither the publisher nor the authors or the editors give a warranty, expressed or implied, with respect to the material contained herein or for any errors or omissions that may have been made. The publisher remains neutral with regard to jurisdictional claims in published maps and institutional affiliations.

This Springer imprint is published by the registered company Springer Nature Switzerland AG
The registered company address is: Gewerbestrasse 11, 6330 Cham, Switzerland

Organization

Program Committee Chairs

Marco Ceccarelli	University of Rome Tor Vergata, Department of Industrial Engineering, Roma, Italy
Marco Paoloni	ENEA, Robotics and Artificial Intelligence Laboratory, S. Maria di Galeria (Roma), Italy

Program Committee Members

Jorge Enrique Araque Isidro	University of Rome Tor Vergata, Industrial Engineering, Rome, Italy
Daniele Cafolla	IRCCS Neuromed, Biomechatronics Unit, Pozzilli, Italy
Giacomo Cupertino	ENEA, Robotics and Artificial Intelligence Laboratory, Santa Maria di Galeria (Roma), Italy
Matteo Russo	University of Rome Tor Vergata, Roma, Italia

Preface

This proceedings book aims to present advances in the field of Space Robotics as related to the topics in the TORVEASTRO project, from theoretical contributions to experimental and practical applications. The goal of the TORVEASTRO project is the development of a new service robot that could assist astronauts in their Extra-Vehicular-Activity (EVA) and can work autonomously or supervised in orbiting space stations.

The book contains the contributions that have been accepted for presentation at the TORVEASTRO 2023: International Conference on Robots for Space Applications in Orbital Stations at the University of Rome Tor Vergata in Rome, Italy, during 20–21 April 2023.

The aim of the conference is to bring together invited researchers, industry professionals and students from the broad range of disciplines referring to Space Robotics, in an intimate, collegial and stimulating environment.

These contributions were accepted after a blind peer-review process, giving an overview of the activity and results of the TORVEASTRO project with additional papers on topics related to the main goals of the project.

We wish to express our sincere gratitude to the authors and reviewers for their valuable scientific contributions and useful feedback in due time to ensure a high scientific quality of this proceedings volume. We also thank Lazioinnova for supporting the TORVEASTRO project through the grant POR214-2020-A0375-2020-36517. The editors are also grateful to their institutions of affiliation, namely the University of Rome Tor Vergata and ENEA, for their support for the conference organization.

Finally, we thank all the staff members at Springer for their valuable editorial contributions.

Rome, Italy
January 2023

Prof. Marco Ceccarelli
Prof. Loredana Santo
Dr. Marco Paoloni
Ing. Giacomo Cupertino

Contents

TORVEASTRO Project: Plans and Achievements

Marco Ceccarelli(✉)

Laboratory of Robot Mechatronics, University of Rome Tor Vergata, 00133 Rome, Italy
marco.ceccarelli@uniroma2.it

Abstract. TORVEASTRO project is outlined with the planned activity on problems for a design of a new space service robot for out space surveillance and maintenance of the orbital station structures. The proposal is based on a conceptual solution with three leg/arm limbs whose operation is actuated by cables with actuators inside the robot central body. The two-year project is successful worked out with design results of a novel solution whose operation is tested successfully in lab experiments with here-in summarized main results including the plans for the last period of the project.

Keywords: Space Robotics · Project planning · Robot Design · Prototype · Testing · TORVEASTRO robot

1 Introduction

In hostile environments, such as outer space, it is convenient to use robots instead of humans both for astronaut safety and operation efficiency. Therefore, robots increasingly play a key role in on-orbit servicing operations. TORVEATRO project deals with the development of a astronaut robot for servicing in orbital stations, [1].

For more than fifteen years, great attention has been paid to service robotics for the development of new systems in addition to those already developed for medical care, space exploration, demining operations, surveillance, entertainment, museum guidance. The growing activity of space exploration sets many future goals and needs. Among these, in Earth's orbit, the growth of debris, resulting from decommissioned spacecraft launched in previous times is one of the main concerns when considering the efficient operation of orbital stations. Therefore, on-orbit operations will be required to clean work zones and maintain spacecraft and orbital stations in efficient service. In order to avoid costly and risky tasks manually handled by astronauts, space/astronaut robots could perform the tasks for those service in space. There is already a considerable literature on service robots referring not just technical issues. Service robotics in space is considered one of the most promising approaches for maintaining systems in orbit with missions such as docking, mooring, refueling, repairing, upgrading, transporting, salvaging, removing orbital debris. The characteristics of a space environment are related to low air pressure, large temperature variations, high radiation and microgravity. Humans, in order to work in space, require a life support system that is large, bulky and complex.

© The Author(s), under exclusive license to Springer Nature Switzerland AG 2023
M. Ceccarelli et al. (Eds.): TORVEASTRO 2023, MMS 130, pp. 1–26, 2023.
https://doi.org/10.1007/978-3-031-28447-2_1

Therefore, the implementation of an environmental control system and a rescue system is necessary, and this requires a considerable material supply. The International Space Station (ISS) is habitable, in low Earth orbit, thanks to a complex structural and management project. Its first component was launched into orbit in 1998, with the first long-term residents in November 2000. The IIS serves as a microgravity and space environment research laboratory where crew members conduct biology experiments, physics, astronomy, meteorology and other fields. IIS is equipped with handrails placed on the outside to help astronauts move in EVA (Extra-Vehicular-Activity) operations. The design of a space robot must meet the requirements of space station operations but, at the same time, its mass and volume should be as small as possible. With the development of space technology, more and more complex space tasks will be carried out in the future, and service robots will increasingly play a key role in on-orbit service operations.

In April 1993, for the first time in the history of spaceflight, a small multisensory robot performed a variety of tasks aboard a spacecraft in different operating modes. Such activities are often carried out today, with remotely controlled operations pre-programmed by astronauts using video tele-monitoring. One currently operating astronaut robot is Robonaut (NASA, National Aeronautics and Space Administration), [2], which was developed at NASA's Lyndon B. Johnson Space Center (JSC) in Houston, Texas. The latest version of Robonaut is a robotic torso designed to assist crew in EVA activities with variety of tools also used by astronauts. Another astronaut robot is ESA's Justin, [3], with four-fingered hands in a light arm making him an ideal operating platform for sensitive manipulations. Kirobo, [4], is the first Japanese astronaut robot (JAXA, Japan Aerospace eXploration Agency) which was developed by the University of Tokyo and Tomotaka Takahashi in collaboration with Koichi Wakata. The European Space Agency (ESA) and the Italian Space Agency (ASI) are currently developing a space robotic system called JERICO, [5], which should be installed in the SPEKTR module of the MIR station. JERICO consists of a seven-axis robot with a sensorized end-effector. Microgravity facilitates movement but the stability of mechanisms can be very sensitive to small vibrations. All the above astronaut robots show successful results as well as additional problems to be solved for a future more efficient application of them in orbital stations, mainly in servicing in surveillance, monitoring and repairing of the outer space structures of an orbital station. Other examples of astronaut robots can be identified in Canadarm, [6], and ESA EuroBot, [7]. Canadarm was designed by Canadian Space Agency for payload handling, support in extravehicular activity, and station assembly. A second version Canadarm2 is also used for the service and maintenance of space stations, transport of payloads, as well as for berthing/unberthing of the orbit. The ESA EuroBot robot is designed for ESA by a consortium led by Thales Alenia Space with a central body structure with three identical arms, each with 7 joints. Each arm is equipped with a camera and an end effector hand, which is capable of grabbing EVA handrails.

The final objective of the TORVEASTRO project, [1], financed by Lazioinnova within research program of the Latium Region in Italy, is the development and design of a new astronaut robot for EVA (Extra-Vehicular-Activity) service activities on orbital stations for maintenance and monitoring operations outside the orbital station structures. The validation of a demonstrator prototype is one of the final deliveries of the project, as results of the research and design activities. The subjects in the project activity include the design of new mechanisms and new solutions for structure and operation of a service astronaut robot for orbital stations while maintaining the peculiarities of lightness and versatility indispensable in a space environment.

2 Requirements for Space Robots Servicing in Orbital Stations

A hostile environment such as space has several peculiarities very different from those in so-called service applications on the earth. Radiation can be dangerous for actuators, controllers, drivers, sensors and electronic devices in general. In addition, space is characterized by sharp changes in temperature. In orbital stations, the energy source is very limited and therefore the energy consumption must be kept as low as possible. It is necessary to consider that it is difficult to transport objects so to make it easy to transport the robot to the space station, it should be as light and small as possible, as underlined in the large bibliography of reference. In this context, space operations of astronauts are not only dangerous but also expensive. Therefore, service robots can be used as assistants or substitutes of astronauts more and more frequently to reduce risk and cost. Furthermore, space robots are more efficient than humans in terms of carrying load, positioning accuracy, and adaptation to work environments over long periods. Even science experiments can be performed by robots so that plans are under consideration for reducing the human stay in orbital stations.

The following are data and aspects that can be considered as requirements and/or constraints for the design and operation of an astronaut robot in orbital station running at about 27.400 km/h at about 400 km from the earth surface for outer space servicing, [8]:

- Presence of ionizing radiation (flux of cosmic rays)
- Accumulation of electric charges on metal parts with the formation of electric arcs
- Space debris produced by old missions
- Temperature variation from −50 °C to 150 °C
- No gravity condition (microgravity at about 0.1 g)
- Low Power consumption (100 W)
- Long operational life (6 h)
- Autonomous operation
- Resist high vacuum in space
- Small size (diameter of 260mm–180mm)
- Equipped with variety of sensors
- Camera for video navigation
- Wireless network connections
- Motion capability among installed handrails
- Operating basic functions for positioning and assembly of station components
- Maintenance and replacement of faulty components of the outside structures

All the above can be summarized in the flow chart in Fig. 1, where general features of an astronaut robot servicing in outer space of the orbital stations are indicated as main requirements for the planned TORVEASTRO project.

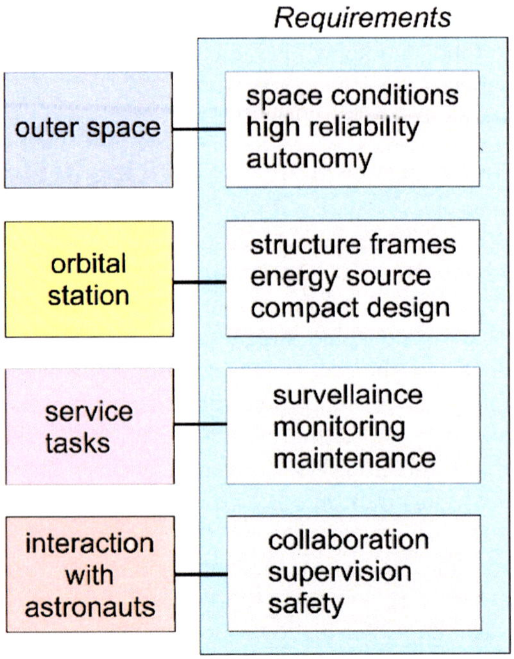

Fig. 1. A scheme of main requirements for the aspects in TORVEASTRO project

3 TORVEASTRO Design in Project Plans

The TORVEASTRO space robot is designed with a cylindrical body with three limbs, each of which is composed of three links and an end-effector. The symmetrical assembly of the limbs with leg or arm operation capability makes them interchangeable also with the possibility of adopting multifunctional end-effectors. The conceptual design is shown in Fig. 2. In practice, the robot behaves like a chameleon with two legs and a tail. Each of these three limbs (all of the same design and operation features) can be used independently or in combination to perform a job and/or to remain attached to the outer structure of an orbital station. The design idea is based on a new structure conceived as a result of experiences for the development of robotic assistance solutions for astronauts with a conceptual idea deposited in an Italian patent application and following initial laboratory experiences with a first prototype as described in (Samani and Ceccarelli 2019).

Figure 2 shows a conceptual diagram of the three-limbed astronaut robot in working configuration. Figure 3 illustrates a conceptual diagram of the three-limbed astronaut robot in the minimum-volume home configuration. Figure 4 shows the structure of a limb with indication of the components in a sagittal view. Figure 5 shows the structure of a limb with indication of the components in a transverse view.

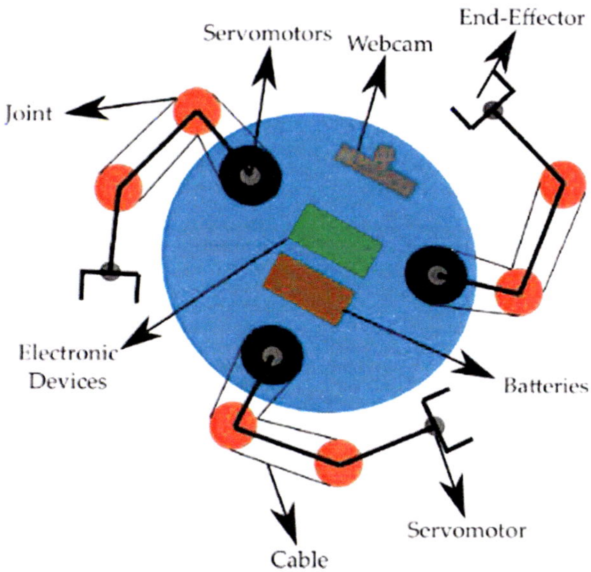

Fig. 2. A conceptual scheme of the design of TORVEASTRO astronaut robot

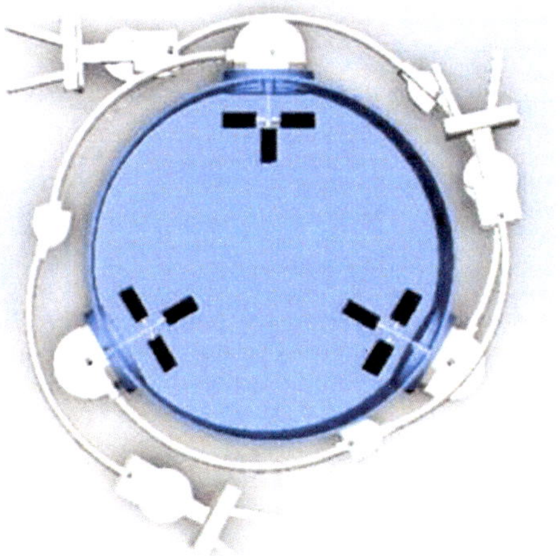

Fig. 3. A scheme of home configuration of TORVEASTRO astronaut robot

With reference to Figs. 2 and 3, the astronaut robot consists of a central body with a cylindrical shape and three limbs with the same shape and dimensions installed symmetrically on the central body.

With reference to Figs. 3 and 4, each limb is composed of a curved arm rod (5), a curved forearm rod (6), a foot/hand end-effector with two fingers (7), an actuator (8) for the shoulder joint, an actuator (9) for the arm joint, an actuator (10) for the transmission in the forearm with pulley (11) of elbow (13) and belt (12), and an actuator (14) for the end-effector. In particular, the belt drive for the stick consists of a belt (12) which runs from the actuator (10) to the elbow pulley (11) inside a tube which is curved in shape like the rod of the arm (5). Furthermore, the actuators (8, 9, 10) of each limb are sized and assembled inside the central body in an actuation unit with a compact shape which allows a home configuration at minimum useful volume in the phases of launch and inactive parking of the astronaut robot.

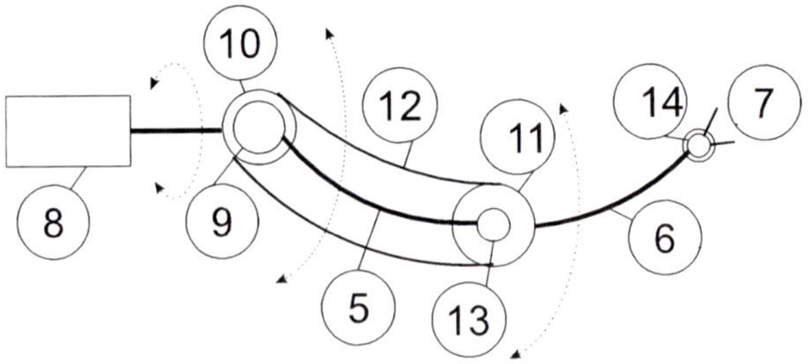

Fig. 4. A conceptual scheme of the structure of a limb with indication of the components in a sagittal view

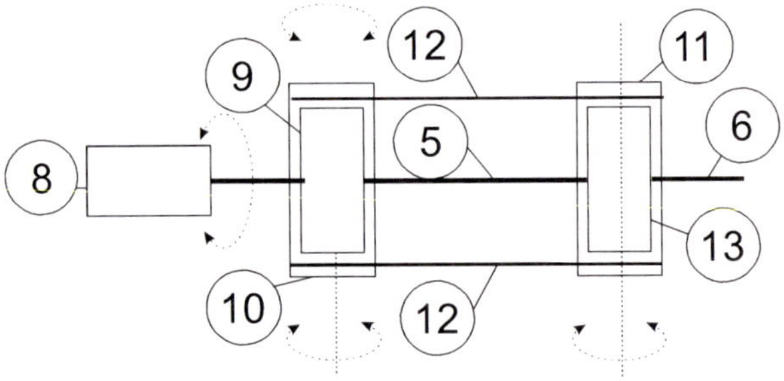

Fig. 5. A conceptual scheme of the structure of a limb with indication of the components in a top view

The main advantage of the three-limbed astronaut robot consists in its compact structure with a constructive symmetry that allows the use of the limbs indifferently in locomotion and manipulation operations outside a space orbital station thanks also to the end-effector solutions with operation by foot or by hand respectively. The end-effector of each limb of the astronaut robot is equipped with a two-finger gripper which, although intended for a foot function, is designed to grip the handles in the external surfaces of the spacecraft or to move by gripping the handrail rods of the external structures. The new three-limb astronaut robot structure can find fields of use mainly, but not only, in external service actions on space orbital stations with the possibility of carrying out inspection and maintenance operations thanks to the locomotion and manipulation capabilities using the three limbs even differently.

Figure 6 summarizes the design and functional structure as conceived for the project activities with an indication of the main aspects that will be developed in the definition of the expected results in terms of the prototype and specific solutions that satisfy the requirements outlined above. It should be noted the centrality of the mechanical design aspects referring to the structure with integration with sensors and actuators, despite the importance of the interfaces and control necessary to achieve the performance for the service purposes of the astronaut robot in space outside orbital stations. Complementary but still essential are the characteristics referring to the ability to communicate data and commands in the programming and execution of service tasks with a monitoring useful for control but also for supervision by the astronauts yet.

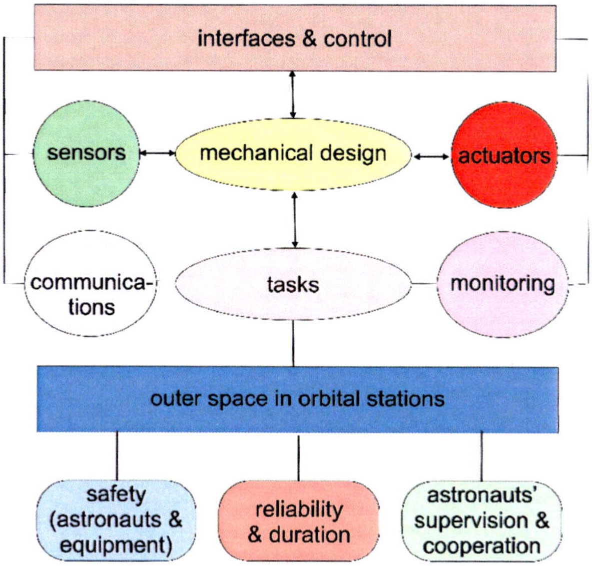

Fig. 6. A conceptual scheme of the design structure of TORVEASTRO astronaut robot with main issues in the space service operation

Figure 6 points out the necessary and characterizing aspects of the development of a service astronaut robot for outdoor environments in terms of safety of the astronauts in collaborative operations and of the same equipment with which the astronaut robot should interact, reliability and adequate duration of functioning in operations in outer space with characteristics necessary for environmental and functional severity, and of operational capabilities that can be carried out with a supervision and cooperation of the astronauts present in the orbital station as well as with remote assistance from the station on earth.

4 Structure and Plans of TORVEASTRO Project

The structure and plans for the activities in the TORVEASTRO project, [1], are summarized in Figs. 7 and 8 and Tables 1, 2, 3 and 4. The TORVEASTRO project is programmed with activities organized in five WPs as in Fig. 7 during three phases as described below with a program schedule summarized in the GANNT in Fig. 8.

The main activities refer to an adequate definition of the requirements and expectations of the system being designed as summarized in Figs. 1 and 6, in terms of the results of the system design and in particular of the control with adequate interfaces and sensors, and for the construction of a demonstrator prototype for demos that can clarify the feasibility and characteristics of the new astronaut robot solution.

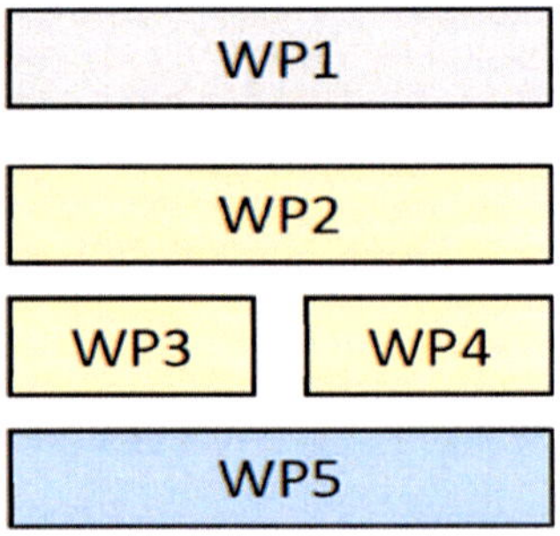

Fig. 7. The structure of the organization for the TOVESATRO project in Work Packages, Table 1

month/WP	1	2	3	4	5	6	7	8	9	10	11	12
WP 1	M1.1	D1.1										M1.2 D1.2
WP 2		D2.2			M2.2	M2.1 D2.1						
WP 3												
WP 4												
WP 5												

month/WP	13	14	15	16	17	18	19	20	21	22	23	24
WP 1	D1.3										D1.3	M1.2 D1.2
WP 2												
WP 3			M3.1					M3.2		D3.1		
WP 4			M4.1					M4.2		D4.1		
WP 5										M5.1 D5.1		

Fig. 8. The GANNT activity planning for the TOVESATRO project, Tables 2, 3 and 4

The activity phases of the TORVEASTRO project in the expected duration of 24 months are planned with a time sequence as described in the GANNT scheme in Fig. 8 with the following activities:

PHASE 1 - from month 1 to month 6: start of activities also with a specific kick-off meeting; updating and research of the problems and requirements required for an astronaut robot in the external maintenance of orbital stations with first definition of

Table 1. Work Packages of TORVEASTRO project

WP No	Work Package	Leader unit	Man months	Starting month	Ending month
1	Coordination, Administration & Dissemination	TORVERGATA	2	1	24
2	Requirements, fields of application & attraction of stakeholders	ENEA	4	1	6
3	Functional design & performance analysis	TORVERGATA	18	3	22
4	Control design, interfaces & sensors	ENEA	14	3	22
5	Demonstrator prototype & demonstration tests	TORVERGATA	10	14	24

Table 2. List of milestones for activity program of TORVEASTRO project

M No	Milestone	WP	Month	Verification result
1.1	Kick-off Meeting	1	1	Worked out
1.2	Report of activity	1	12, 24	Completed & approved
1.3	Workshop – Conference	1	12, 24	Worked out
2.1	Definition of project requirements	2	6	Completed & approved
2.2	Information to potential stakeholders	2	5	Worked out
3.1	Definition of project design solution	3	15	Completed & approved
3.2	Definition of functional performance of design solution	3	20	Completed & approved
4.1	Definition of project design solution	4	15	Completed & approved
4.2	Definition of functional performance of design solution	4	20	Completed & approved
5.1	Construction of demonstrative prototype	5	22	Completed & validated

Table 3. List of deliverables for activity program of TORVEASTRO project

No	Deliverable	WP	Results type	Month
1.1	Report of Kick-off Meeting	1	Report	2
1.2	Report of progress	1	Report	12,24
1.3	Workshop and Conference	1	Announcement and Proceedings	13,23
2.1	Design requirements	2	Report	6
2.2	Flyer of the project	2	Flyer	2
3.1	General executive design	3	Report and construction drawings	22
4.1	Executive design of control	4	Report and construction instructions	22
5.1	Demonstrative prototype	5	Prototype	22

Table 4. An analysis of risk for activity program of TORVEASTRO project

Description and level of risk	WP	Action for risk reduction
High: difficulties for partner meetings, workshop and conferences due to the COVID pandemic	WP 1 – WP 5	Organization of tele-meetings and request for special permits
Medium: uncertainties in the identification of requirements in applications on orbital stations	WP2	Interaction with stakeholders and possibly with ASI and ESA
Low: determination of unconvincing solutions in terms of efficiency and satisfaction of requirements	WP3	Reiteration of the design process and performance analysis, even with pre-tests
Low: determination of unconvincing solutions in terms of efficiency and satisfaction of requirements	WP4	Reiteration of the design and identification process of interfaces and sensors, even with pre-tests
Medium: non-completion of the construction of the demonstrator prototype due to difficulties in delivery of commercial components	WP5	Anticipation of scheduling and request for delivery of material by express courier

design solutions both for the entire system and its components in terms of suitable capabilities in the space environment and for a low-cost demonstrator for experimental validations with Earth gravity laboratory demos.

PHASE 2 from month 6 to month 18: detailed design of the entire system and its components both with the characteristics suitable for an environment in space outside the orbital stations and with low-cost solutions for an Earth gravity laboratory demonstrator; start of the construction of the low-cost demonstrator for earth gravity operation and experimental validation tests for the definition of final demos; at the end of the first year it is planned to organize a workshop with limited participation of potential stakeholders for the development of the astronaut robot and for sharing the progress in the project with interested stakeholders.

PHASE 3 from month 18 to month 24: finalization of the design solutions suitable for the environment outside the orbital stations and for the laboratory demonstrator operating under earth gravity; construction of the demonstrator prototype with related functional verification activities; presentation of the demonstrator prototype with demos in the presence of all project partners and invited stakeholders; at the end of the project at month 24 organization of a congress also with international participation for the presentation of the project results and also presentation of the demonstrator prototype with demos qualifying the obtained results for the functionality of the proposed astronaut robot.

The roles of the personnel expected to carry out the project activities can be summarized as follows:

Tor Vergata unit: Prof. Marco Ceccarelli has the role of coordinating the entire project but will also carry out activities contributing to the development of the design of the entire system, especially with attention to the mechanisms and structural solutions of the astronaut robot, with guidance and supervision of the related research fellows involved in these activities; Prof. Loredana Santo will have a supervisory coordination role of the design activities especially with regard to the structure and the choice of materials for the construction project with guidance and supervision of the related research fellows involved in these activities; a research fellow to be defined with the relative recruitment within the first 4 months will be dedicated to the structural design and kinematics aspects for the astronaut robot with related simulation and performance analysis activities also contributing to the development and construction of the demonstrator prototype; a second research fellow to be defined before the end of the first year with the relative recruitment will be dedicated to the construction and experimental validation of the demonstrator prototype in laboratory activities and technical-scientific dissemination. The members of the unit at University of Rome Tor Vergata (in short Tor Vergata unit) are Prof. Marco Ceccarelli, Prof. Loredana Santo and as research fellows Dr. Daniele Cafolla and Mr. Jorge Araque Isidro, Fig. 9.

ENEA unit: Dr. Marco Paoloni has a coordination role of the unit and will carry out activities within the WPs, especially WPs 2, 4, 5 relating to the design and implementation of control and systems of sensors and interfaces described above also with related experimental laboratory tests, with the guidance and supervision of the related research fellow involved in these activities; Ing. Giacomo Cupertino will have a role in the design and characterization activities with verification of the performance and characteristics of the control and sensor systems and interfaces described above also with related experimental laboratory tests, with guidance and supervision of the relevant research fellow involved in these activities; a research fellow to be defined within the end of the first year with the relative recruitment will be dedicated to the design of the control and sensor systems and interfaces described above also with the relative experimental laboratory tests, with low-cost solutions for the demonstrator prototype and with solutions suitable for the external environment in orbital stations. The members of the unit at ENEA (in short ENEA unit) are Dr. Marco Paoloni and Ing. Giacomo Cupertino, Fig. 9.

Fig. 9. The members of the units for TOVESATRO project at a meeting on 19 December 2022 with a first full prototype: (from left to right): Prof. Marco Ceccarelli, Prof. Loredana Santo, Ing. Giacomo Cupertino, Ing. Jorge Araque Isidro, Dr. Daniele Cafolla, and Dr. Marco Paoloni

With reference to Fig. 7 and Table 1, the activities are distributed in 5 WPs (Work Packages) with activities that are planned according to the phases described above, with synergy and collaboration between the partners participating in the project while having one of the partners as the main reference and specific coordinator of each WP.

WP 1 is aimed at the coordination, administration and dissemination of the project results under the main responsibility of the Tor Vergata unit with activities during the entire duration of the project, also including aspects of control and correction of the activities of the other WPs in accordance with the proposal of the project and any corrections during the progress of the research. These activities of verification and adjustment of the activities of the other WPs by WP 1 will in any case be shared with all the project partners in appropriate periodic meetings.

WP 2 is dedicated under the main responsibility of the ENEA unit to the definition of the requirements and fields of use of the astronaut robot with an updated review of the current state of needs in the orbital stations and in the available solutions. This WP will also work for the initial project visibility with information to companies in the aeronautical and aerospace sectors, especially the regional ones, with possible initiatives in meetings and production of information material.

In WP 3 the activity is concentrated to the design of the proposed astronaut robot under the main responsibility of the Tor Vergata unit with the aim to achieve characteristics suitable for the outer space environment of the orbital stations and low-cost solutions for the laboratory demonstrator prototype, as due to the limited budget available for material purchase. In particular, the design will be focused on the structural solutions with proper kinematics to have the limbs with arm and leg functionality in such a way that the actuating servomotors can be placed inside the cylindrical volume whose goal is also the structural and functional safeguard of the hardware and software. In addition, activity will be carried out for the design of the functionality in terms of mobility and programmability of suitable movements using automatic solutions and/or remotely coordinated by astronauts via telemonitoring. Mechanism designs are expected to be used in order to ensure transmission of motion and forces to the respective articular joints of the limbs for operations of handling objects and locomotion of the astronaut robot along the external structures of the orbital stations. The innovative design can be recognized in the design solution with the three limbs with leg and arm operation capability both for the interchangeability of operations and for the configuration flexibility in the various monitoring and maintenance operations outside the orbital stations. Particular care will be taken in the design to ensure that the astronaut robot in its home configuration can have a minimum size as shown in the conceptual diagrams of Fig. 3. The central body with a cylindrical geometry will be examined in terms of volume suitable to accommodate not only the actuations with the servomotors and the hardware necessary for operation, but also to have a suitable volume for any further storage of other instrumentation. Furthermore, specific attention will be addressed to the problem of choosing the material that can meet the requirements not only of robustness but also of protection against radiation and thermal stress. In particular, the design will be carried out along two lines, i.e. considering a final solution in a space environment outside the orbital stations and a solution for the low-cost demonstrator prototype for laboratory presentations under terrestrial gravity. In both solutions the peculiar characteristics of structural solutions and mechanisms will be considered specific components will be used for the construction and operation of the demonstrator prototype. Significant activity is planned for the analysis of performance with multi-body dynamic simulations useful both for a performance characterization and for the design optimization of the solutions.

WP 4 is planned with specific activity on control design under the responsibility of the ENEA unit with interactions with WPs 3 and 5 for the final solution of the system, and with WP2 for the necessary scheduled references. The control system model for TORVEASTRO robot is structured on three levels, hierarchically distinguished by the complexity of the actions to which they are dedicated. The first level, which can be defined as "upper level", concerns the more complex actions necessary to complete a work mission. For example, if the mission consists in the maintenance of a component of the station, the upper level of the control system would have to identify the movements along the handrails of the ISS that the robot must perform to reach the component, the correct sequence of maneuvers necessary to the maintenance operation (screwing nuts, removing panels, and so on) and the subsequent movements to reach a position where it can be recovered (which does not necessarily have to be the same from which it started). It will therefore be a question of adapting to the robot sequences of operations that are generally already well codified to be performed by the astronauts. The second level, as an "intermediate level", analyzes the actions to break them down into still relatively complex operations. For example, this level will deal with how the robot can move along a certain handrail. For this operation, while the robot is firmly clinging to the handrail via an end-effector, the second level of the control system will move the three limbs in order to bring another end-effector to a new position to cling to and to free then the taking of the first. It should be noted that these operations could sometimes be performed in different ways as for example, if the sequence described were performed using only two limbs repeatedly, the robot would "walk" along the handrail, while if it were performed cyclically using all three limbs, the robot would "roll". This level of the control system would therefore also be responsible for identifying the optimal solution according to the activity controlled by the first level (for example it is reasonable to expect that rolling is faster than walking, but the latter leaves free a limb that could be used to carry a load). The third level refers to the actions of specific single components, in agreement with the superior two levels. The components for controlling the servomotors could be microcontrollers for which ENEA will take care of the programming. A similar solution could also be envisaged for the acquisition of images from the cameras. Checks on the correct functioning of this level of the control system can be carried out by creating a test bench that can be equipped with a camera and a complete limb on which to carry out the experimental and trial tests. Once the lower level has been consolidated, it will be possible to move on to the implementation of the intermediate level with activities for the development of software such as, for example, the direct and inverse kinematics of the limbs and of the robot as a whole, the processing of images from cameras for object recognition and photogrammetry measurements. This software will run on one or more CPUs that will be selected to ensure the proper execution speed required by the robot controller and the integration between CPUs and microcontrollers.

WP 5 is specifically programmed for the development, construction and functionality check of a low-cost demonstrator prototype, considering the limited budget available for the purchase of equipment. The demonstrator purpose is not only to validate the design solutions defined both at a general and specific level but also to present a demonstrator with relative demos to potential stakeholders, soliciting further collaborations and future developments also in further projects in both national and European funding programs. In this WP coordinated by the Tor Vergata unit, strong synergies between all the project participants and even collaborations with potential stakeholders in the space sector of the region are expected. This international congress scheduled at the end of the project foresees a significant activity by WP5 precisely with the presentation of the demonstrator prototype to the technical-scientific community, with the possibility of further dissemination even to the wider public, not only in the Lazio Region.

Tables 2 and 3 list the Milestones and Deliverables that characterize the activities described above with expected results from the project activities in accordance with the outlined technical-scientific program. The final result is expressed in the form of the demonstrative laboratory prototype of the TORVEASTRO astronaut robot. The timing of the Milestones and Deliverables is reported in the same GANNT of Fig. 8 to better characterize the evolution of the project activities in its phases and in its work packages WPs.

The Milestones in Table 2 refer to the main results expected in the context of activity development with indications of the individual WPs and the temporal deadline.

Table 3 shows the Deliverables in terms of the documentation that is intended to be produced as proof of the results obtained in the individual WPs in accordance with the aims of the project and the expected results by using reports, brochures and technical-scientific publications from congress presentations and journal submissions.

In addition, potential risks of failure of the activities were analyzed in the main risk aspects as summarized in Table 4 with reference to the main activities of the individual WPs and indicating measures for risk mitigation and to avoid those failures.

5 Expected Results and Problems

The project plans to develop a demonstrator that can clarify the interest and the practical feasibility of using the proposed innovative astronaut robot structure as the main result of the project. To reach the definition of the demonstrator prototype and to complement the value of it the following main results are expected as planned in the proposed project plan, still referring to Milestones and Deliverables in Tables 2 and 3, in term of project activity as:

- Completion of the WPs activities following the planned GANNT outline
- Recruiting of research fellows for each project unit
- Dissemination activity towards potential stakeholders, technical-scientific communities, and general public

and in term of technical-scientific achievements as:

- Definition of design and operation requirements
- Identification of solutions for in-orbit and on-earth astronaut robot TORVEASTRO
- Design of a prototype demonstrator of on-earth astronaut robot TORVEASTRO
- Testing and demo of the prototype demonstrator of robot TORVEASTRO
- Patent releases for the innovative solutions
- Publications in conference proceedings and journal for dissemination of the results

During the development of the activities the following problems have been experienced:

- Delay in first year activities, mainly due to the effects of the COVID-19 pandemic in availability of the infrastructures ta the project unit institutions and in delivery of the needed/requested material and market components
- Difficulties and then failure in recruiting a research fellow for the ENEA project unit, causing delay and reduction of activity of the planned activity of the ENEA unit
- Difficulties in dissemination activities as due to limitations for the fighting against the COVID-19 pandemic

Nevertheless, the project activity has run regularly and continuously also with telemeeting contacts and exchanges of information among the project units. In the last period the activities have been speed up thanks also to the recovering the regular functioning of all the frames, giving also the possibility to regain the delays and to solve the lack of support of the research fellow at ENEA unit with alternative additional efforts by the unit members.

6 Achieved Results

The project activities were carried out regularly and continuously according to the program presented in terms of phases with the activities in the WPs as:

- WP1: Management, Coordination & Dissemination
- WP2: Requirements and business attraction (ended on 15 October 2021)
- WP3: Design and Performance (started om July 16, 2021)
- WP4: Control, interface and sensor design (started on June 16, 2022)
- WP5: Prototyping and Testing (started on August 16, 2022)

In particular, the main carried out activities with results refer to:

- Analysis of project and functional requirements
- Analysis of suitable materials for robotic structures in space
- Modeling and calculations for dimensioning the body and parts of the astronaut robot in gravity and space versions
- Modeling and calculations for the sizing of the body and arm-leg limbs for the demonstrator prototype
- Modeling and calculations for kinematic control algorithms
- CAD modeling and design of parts with functional material
- Construction of prototypes by using market components and 3D printing with PLA
- Experimental design and testing of prototypes for functional validations
- Planning and carrying out of experiments with related data acquisition for functional performance characterization
- Elaboration of publications for the dissemination of the results and of the project

The main objectives achieved so far, in accordance with the project planning also in terms of milestones and deliverables in Tables 2 and 3, can be considered in:

- Identification of project and functional requirements
- Sizing of the body and parts of the astronaut robot in gravity and space versions
- Sizing of the body and the arm-leg limbs for the demonstrator prototype
- Formulation of kinematic control algorithms
- Construction and Assembly of parts for prototypes of limbs and full robot
- Experimental results from testing with limbs and full robot
- Elaborations of publications for the dissemination of the results and of the project

The results obtained so far with a very satisfactory progress to the final goals can be indicated from investigation and procedures as:

- Definition of a new structure with constructive and functional details
- Modelling and design solutions for the central body, leg-arm limb, cable limb-actuating system, sensorization, and control unit
- Definition of an innovative solution of cable-driven mechanism joint presented for patent release, [9].

- Identification of market components
- Design and of constructions of parts by 3D printing
- Design of an innovative actuation unit with 3 actuators for each leg-arm limb

and in terms of built prototypes experienced with lab testing as:

- Testbeds with prototypes for leg-arm limb, actuation unit with 3 actuators, cable limb-actuating system, cable-driven mechanism joint, motion-controlled limb, and full demonstrator prototype.

Specific actions were taken for dissemination of the project and its results with:

- Contacts to potential stakeholders in the Lazio region, even by email
- Elaboration and posting of project flyer and poster
- Communications to news media with press releases and article publications in newspapers, [10]
- Organization of a regional Workshop at the end of the first year in April 2022
- Organization of an International Conference TOREVASTRO 29023 at the end of the project in April 2023
- Participation to an international competition of projects in aerospace by the Chinese Space Agency for which the team of Tor Vergata has been selected among finalists for the category 'Innovative Design of Space Payload Experiment' with final judgment in April 2023 (isssp.bit.edu.cn/en/n12/)

The technical-scientific results were also published as reported in references [10–20]. In particular Figs. 10, 11, 12, 13, 14, 15, 16, 17, 18, 19 and 20 summarizes the main results in terms of the constructed prototypes from the design activity including plans with testing for validation and characterization of the performance of TORVEASTRO robot and its components.

Figure 10 shows a conceptual scheme of the structure that is designed for TORVEASTRO robot astronaut with service purposes in the external structures of an orbital station with the characteristics of having three limbs that can function as an arm or leg suitably actuated. The further design peculiarity can be recognized in the fact that all the elements and components sensitive to the extreme conditions in space are housed inside the central body which will be duly designed in terms of structure and materials to safeguard these components. In particular, the scheme indicates the main components (L for limb, M for servomotor, S for sensors on servomotors, T for telecommunication unit and data transfer; B for power supply battery with recharging capacity, C for the control unit, W for webcam or vision systems for remote visual monitoring, and E for sensors and additional equipment), which are expected to be assembled with an appropriate design to achieve an adequate structure and functionality with a modular design solution that can ensure not only high reliability but also convenient interchangeability and complementarity of parts and components of the astronaut robot.

Figure 11 shows the geometric diagram that has been conceived with indications of the design parameters for an accurate sizing of the mechanical structure of the robot both in the limbs with an adequate shape to ensure the home configuration with the smallest possible encumbrance and in the central body.

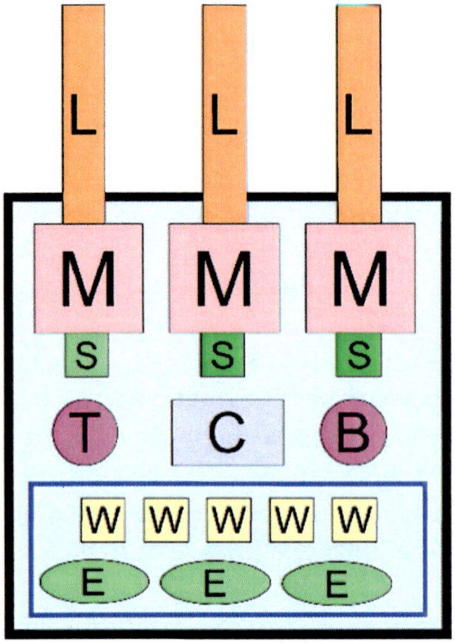

Fig. 10. The conceptual design of the TORVEASTRO service astronaut robot. (L for limb, M for servomotor, S for sensors on servomotors, T for telecommunication unit and data transfer; B for power supply battery with recharging capacity, C for control unit, W for webcam or vision systems for remote visual monitoring, and E for sensors and additional equipment)

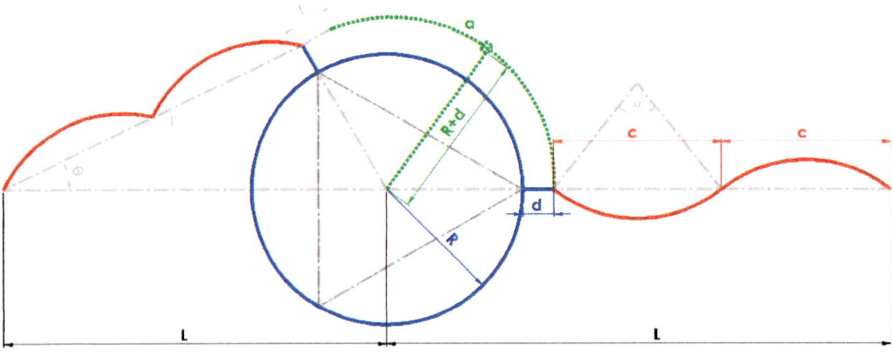

Fig. 11. A geometric scheme for sizing TORVEASTRO design with design parameters

The diagram in Fig. 12 represents the design solution of the central body with the arrangement of the components defined for the on-earth demonstrator prototype consisting of the three limb actuation units (in blue), the control unit consisting of the Raspberry mini controller (in green), and the rechargeable LIPO battery (in orange) as a power source for the whole system.

Figure 13 shows the compact design of the power unit with three servo actuators for one limb with the characteristic of having one direct drive actuator for one shoulder movement and the other two actuators driving cables for a second shoulder movement. And elbow movement respectively. The actuation unit is designed in a compact and modular shape also for the adoption of different types of servo actuators with a house that is easy both for assembly and for power and control wiring.

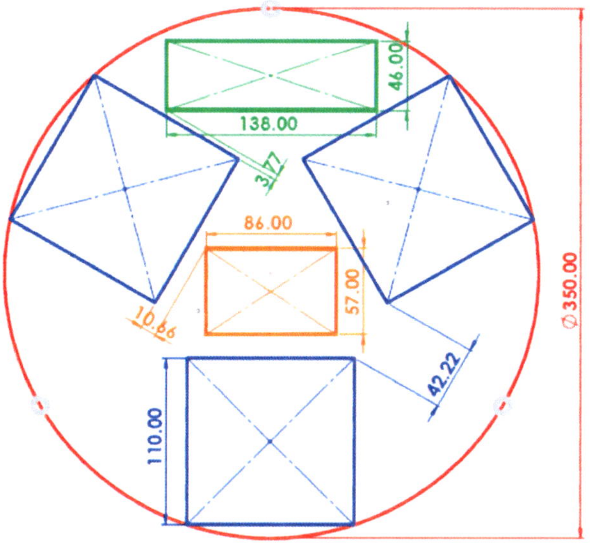

Fig. 12. The model for the central body of the TORVEASTRO demonstrator porotype. (three limb actuation units in blue, Raspberry mini controller in green, and rechargeable LIPO battery in orange)

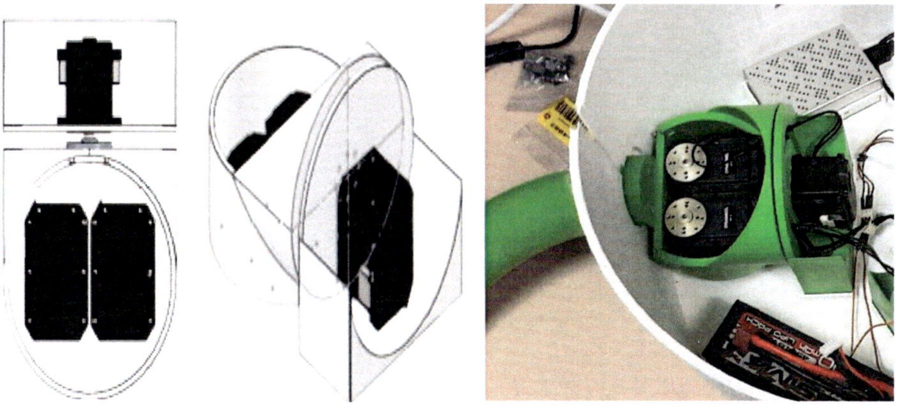

a) b)

Fig. 13. The design of a limb actuation unit: **a** CAD model; **b** lab prototype for testing

Figure 14 shows the mechanical design of the TORVEASTRO robot demonstrator with its characteristics referring to the housing of all the sensitive components in the central body and in the distribution of the three curved limbs for a minimum volume home configuration. In Fig. 1a the dimensions and the movement capabilities of the three limbs are emphasized, while in Fig. 14b the expected shape is shown as coming from the construction of the demonstrator through 3D Printing manufacturing.

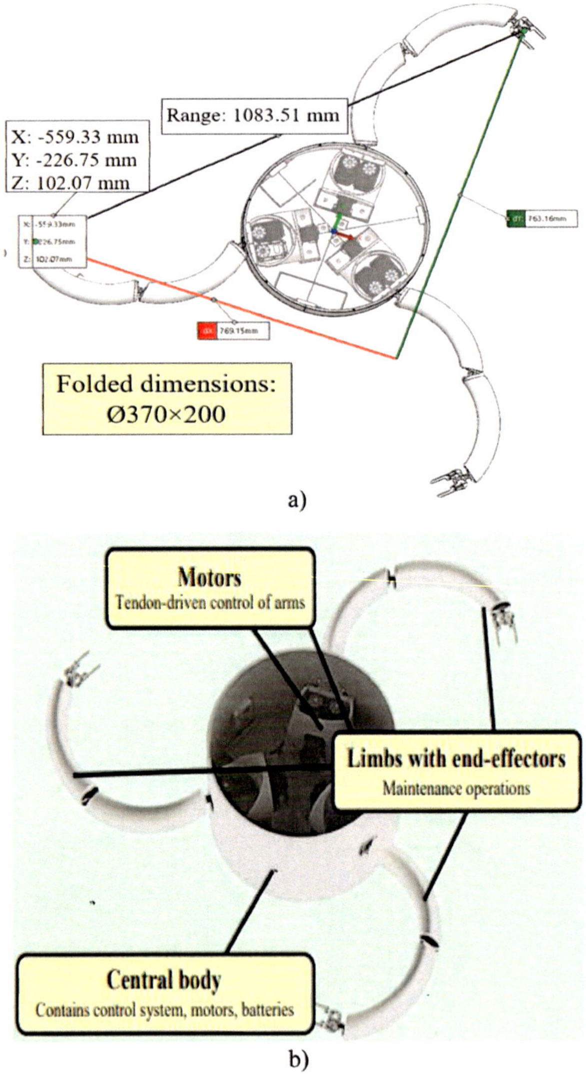

Fig. 14. The mechanical design of TORVEASTRO astronaut robot: (a) a 3D model with sizes; (b) CAD model with main units

Figure 15 shows the demonstrator assembly process with a first analysis of the efficiency and feasibility of the arrangement with the motor units. In Fig. 15a it can be seen how the assembly of the motor unit is easy with a minimum space occupied inside the central body for easy preparation for the wiring. Figure 15b shows the final assembly for a test of simultaneous power supply of the three motor units with computer monitoring for subsequent validation tests and functional characterization.

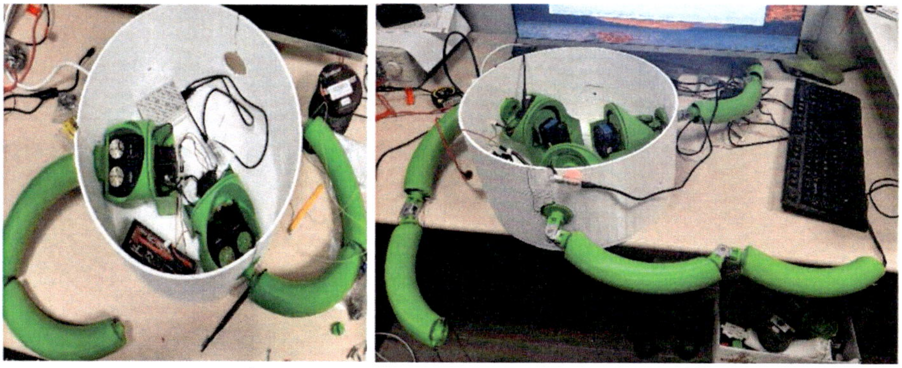

a) b)

Fig. 15. Assembly of the TORVEASTRO demonstrator robot: (a) location of the motor units; (b) lab testing of the

Figures 16 and 17 show a summary of the activities and design solutions in the search for a mechanism-actuated joint with greater efficiency and robustness. The design idea is that of transmitting the actuation motion by cables through a planar mechanism which can effectively and robustly transmit motion and force for the movement of the joint.

Figure 16 shows the first solution investigated with a structure based on a direct slider-crank mechanism with configurations that avoid singularity problems and allow a compact mechanical design. Figure 16a shows the kinematic diagram with the design and functional parameters that allowed an in-depth analysis then corroborated by validation tests and functional characterization with the testbed shown in Fig. 16b purposely made at large scale to emphasize the critical points and peculiarities of the proposed solution.

Figure 17 represents the solution evolved from the one in Fig. 16 by adopting a kinematic inversion of the mechanism in order to increase the efficiency of transmission of motion and force to the joint platform and simplifying the connection to the antagonist actuating cables. The prototype shown in Fig. 17b was defined following adequate analyzes of functional simulation and numerical sizing starting from the kinematic structure of Fig. 17a. However, this solution of actuated joint is still being studied for an efficient application on the demonstrator prototype since the miniaturization of the mechanism is problematic for the housing in the structure of the limb and for the effective actuation with an adequate range of motion of joint.

The problem related to highly compact but robust and efficient joints in the transmission of motion and force is an aspect of a fundamental challenge in defining the mobility of the limbs for a flexible functionality of the astronaut robot TORVEASTRO in service applications both in manipulation and locomotion.

Figure 18 shows a first prototype of the limb actuated with a motor unit via cables for shoulder and elbow movements in a laboratory test for functional validation. The cables are installed through special guides inside the body of the limb segments and are

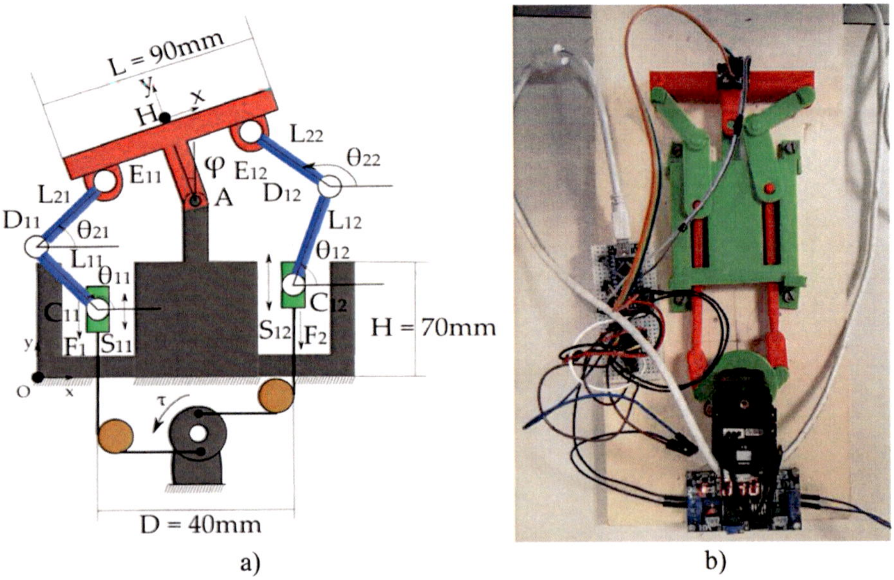

Fig. 16. The model of cable-driven slider-crank mechanism joint for the TORVEASTRO limb: (a) a kinematic model with parameters; (b) a testbed with large-scale prototype

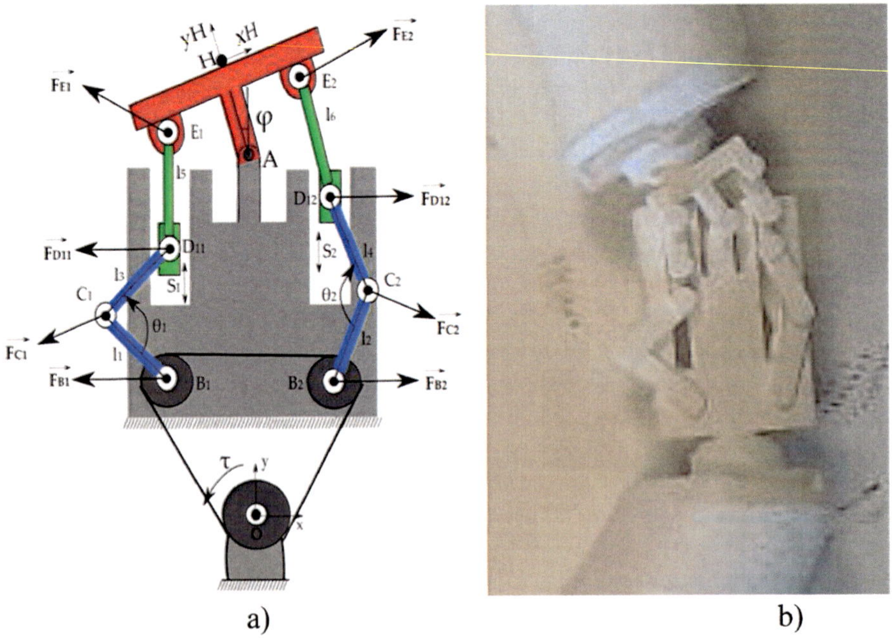

Fig. 17. The model of cable-driven inverted slider-crank mechanism joint for the TORVEASTRO limb: (a) a kinematic model with parameters; (b) a prototype installed on a limb PLA prototype

effective with a tension well below the cable capacity of about 100 N. It can be noted in the reported test the large relative movement of the forearm by the elbow joint with an adequate continuity of motion as indicated by the green trajectory of the center of mass of the forearm body.

Figure 19 shows the complete demonstrator prototype that is assembled with a mass of about 4 kg for a first functional test at the beginning of December 2022 with an experiment layout that suspends the mass of the robot for a movement of the limbs in free space. The dimensions of the demonstrator robot are in accordance with those of the project, referring to the cylindrical body with a diameter of 370 cm and a thickness of 200 cm, and each limb with a circular section of 25 mm radius and a length of 390 cm. A total power of about 10W powered by a LIPO battery is expected by the simultaneous action of the three motor units, whereas each servomotor will provide a maximum torque of 180 N mm well below the nominal capacity. The first validation test was conducted as also indicated in Fig. 20 for a movement from a deactivated robot to the home configuration with the limbs wrapped around the central body with a movement of about 3 s. The demonstrator prototype was built with the structural parts produced by 3D printing in PLA while the functional components were created using commercial products such as Dynamixel MX-64 motors and Raspberry minicontrollers. The sensor equipment with current sensors, IMU, and web cams have also been installed for the next numerical characterizations of the operation in demonstration tests of the performance in the main operations of the robot.

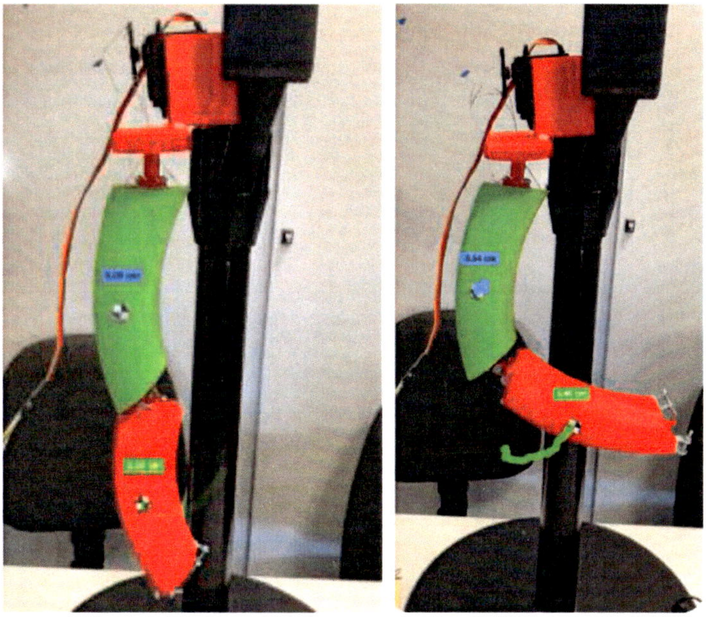

Fig. 18. A snapshot of a test with an actuated limb prototype

The activities of the project will continue until April 2023 and, despite the inevitable problems due to the complication of the project and the experimental activities, it is

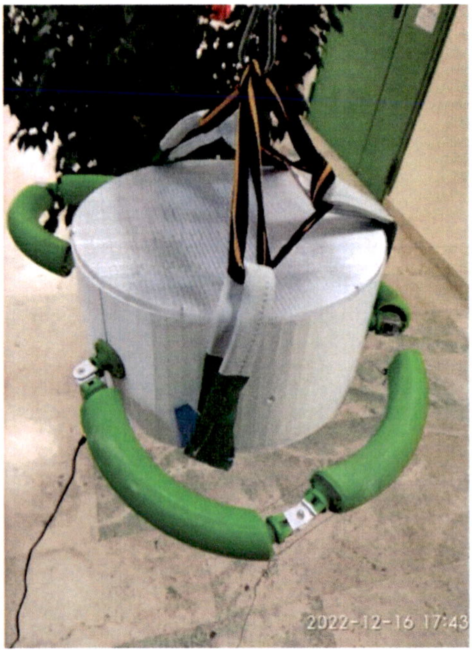

Fig. 19. The full prototype of the TORVEASTRO demonstrator robot

expected to fully achieve the objectives and to define the TORVEASTRO demonstrator astronaut robot with adequate functional and robustness characteristics for qualifying tests of functional characterization and with results that can give the possibility of future extension in designs of real application of the robot on orbital stations.

7 Conclusions

The activities of the TORVEASTRO project, financed in the Lazio regional program POR FESR LAZIO 2014–2020 by Lazioinnova, are presented in the aspects for the development of an astronaut service robot called TORVEASTRO with the innovative solution that is characterized by three cable-actuated leg/arm limbs which are connected to a central body containing all the mechatronic components. Although the design solution is developed with a laboratory prototype for earth gravity operation with functionality for orbital stations, the innovative features are designed for implementations in an astronautical environment with micro gravity, solar/cosmic radiation, and differential heating, with adequate components to the space that can be identified with appropriate elements in a future program with adequate funding.

a) b)

Fig. 20. A lab testing with TORVEASTRO demonstrator robot in Fig. 19: (a) the testing layout with the prototype on the air; (b) in home configuration after motor activation

Acknowledgements. This work has been supported by grant Lazioinnova-POR214–2020-A0375–2020-36517 for the project TORVEASTRO that is gratefully acknowledged.

References

1. Ceccarelli M.: TORVEASTRO—Design and testing of a robot astronaut demonstrator, project no. A0375–2020–36517, program POR FESR LAZIO 2014–2020, University of Rome Tor Vergata, Rome (2019)
2. NASA: Robonaut 2, https://www.nasa.gov/robonaut2. (visited on 2 December 2022)
3. DRL: Rollin' Justin, https://www.dlr.de/rm/en/desktopdefault.aspx/tabid-11427/#gallery/35411. (visited on 2 December 2022)
4. Toyota, KIROBO, https://global.toyota/en/detail/6035471. (visited on 2 December 2022)
5. ESA-ASI, JERICO, https://www.esa.int/esapub/pff/pffv7n2/didov7n2.htm. (visited on 2 December 2022)
6. ASC-CSA, CANADARM, https://www.asc-csa.gc.ca/eng/canadarm/. (visited on 2 December 2022)
7. ESA, ESA-Eurobot, https://www.esa.int/Science_Exploration/Human_and_Robotic_Exploration/Eurobot_Ground_Prototype. (visited on 2 December 2022)
8. Paoloni M.: Report of W2-Torveastro, Enea, Roma (2021)
9. Ceccarelli Marco, Cafolla Daniele, Araque Isidro Jorge Enrique, (Mechanism-driven joint, paptent request no. IT102022000007889, Italy, 1/04/2022

10. Travini, P.: This is how our robots will go in orbit with astronauts, Messaggero newspaper, 12 December 2022, p. 19
11. Samani, F., Ceccarelli, M · Prototype design and testing of TORVEastro, cable-driven astronaut robot. In: Advances in Service and Industrial Robotics—Results of RAAD 2020, Springer Nature Switzerland AG 2020, Cham, pp. 448–455 (2020). https://doi.org/10.1007/978-3-030-48989-2_48
12. Samani, F., Ceccarelli, M.: An experimental characterization of TORVEastro, cable-driven astronaut robot. Robotics **10**(1), 21 (2021). https://doi.org/10.3390/robotics10010021
13. Araque-Isidro, J.E., Cafolla, D., Ceccarelli, M.: Problems and requirements for outer space astronaut service robot. In: Niola, V., Gasparetto, A., Quaglia, G., Carbone, G. (eds.) Advances in Italian Mechanism Science. IFToMM Italy 2022, pp. 603–611. Springer, Cham (2022). https://doi.org/10.1007/978-3-031-10776-4_69
14. Araque Isidro, J.E., Cafolla, D., Ceccarelli, M.: Evaluación experimental del funcionamiento de un prototipo del robot astronauta TORVEASTRO V2, XV Iberoamerican Conference (CIBIM 2022), November 22–24, Madrid, Spain, Paper No. 51, 2022. (in Spanish) . (in print)
15. Paoloni, M., Cupertino, G., Ceccarelli, M.: A kinematic analysis of the TORVEastro astronaut robot arm, the 13th IFToMM Inte. Symposium on Science of Mechanisms and Machines (SYROM 2022), November 17–18, Iasi, Romania, Paper No. 28, Springer, Cham. 2022. (in print)
16. Cafolla, D., Ceccarelli, M.: Dimensional design of Torveastro, a space orbital station service robot. In: Proceedings of TORVEASTRO 2023 conference, 20–21 April 2023, Rome, paper 03, Springer, Cham (2023). (in print)
17. Cafolla, D.: Performance analysis of Torveastro, an outer space service robot In: Proceedings of TORVEASTRO 2023 conference, 20–21 April 2023, Rome, paper 04, Springer, Cham (2023). (in print)
18. Cupertino, G., Cipollini, A., Paoloni, M.: Control design in TORVEastro prototype. In: Proceedings of TORVEASTRO 2023 conference, 20–21 April 2023, Rome, paper 06, Springer, Cham (2023). (in print)
19. Cafolla, D., Araque-Isidro, J.E., Ceccarelli, M.: Design and testing of Torveastro: an outer space service robot. Appl. Sci. **13**(2), 1187 (2023). https://doi.org/10.3390/app13021187
20. Araque Isidro, J.E., Cafolla, D,. Ceccarelli, M.: Performance analysis of a mechanism-driven joint. In: Proceedings of The 6th IFToMM International Conference on Mechanisms, Transmissions, and Applications (MeTrApp 2023), , 24–26 May 2023, Poitiers, paper 04, Springer, Cham (2023). (in print)

Orbiting Space Station Robots: Issues, Challenges and Requirements for a Novel Design

Marco Santoro(✉), Marco Paoloni, and Giacomo Cupertino

ENEA – Italian National Agency for New Technologies, Energy and Sustainable Economic Development – Robotics and Artificial Intelligence (RIA) Lab, 00123 Roma, Italy
{marco.santoro,marco.paoloni,giacomo.cupertino}@enea.it

Abstract. This paper outlines an almost comprehensive description of issues faced by space station robots in order to identify the requirements that a design of a novel service robot has to meet to perform Extra-vehicular activities (EVAs). Existing regulations are reviewed and the main aspects of already operating robots on the International Space Station are underlined. Main criticalities of the Low Earth Orbit environment (radiations, space debris and temperature excursions), as well as those of the launch from the ground to the space station, are described.

Keywords: International Space Station (ISS) · Space robot design · Service robot · Space station environment · TORVEastro robot

1 Introduction

A service robot dedicated to orbiting space stations must necessarily be designed with an unstructured environment in mind for its use; in fact, it is unthinkable to require the organizations operating the stations to modify them to suit the robot to be proposed.

On the other hand, the need to expand the operational capabilities of stations, demanded by the continuous development of technical and scientific programs, requires these same bodies to make available to designers engaged in this type of initiative organic technical documentation, constantly updated, in order to standardize the technical requirements for the selection, application and design of new components and procedures.

For example, NASA conducts a project called the "NASA Technical Standards Program" (NTSP), which is part of the staff of the "Office of the NASA Chief Engineer," charged with keeping the "NASA Technical Standards System" (NTSS) up-to-date. The NTSS administers a website [1] on which NASA's general technical standards can be consulted, which includes both documents that can be consulted publicly for free and others that can be consulted for a fee or are subject to confidentiality clauses.

Similarly, for Europe, the European Space Components Coordination (ESCC), in which ESA, national space agencies and space industries participate, coordinates activities related to the production and use of electrical, electronic and electromechanical components suitable for use in space.

© The Author(s), under exclusive license to Springer Nature Switzerland AG 2023
M. Ceccarelli et al. (Eds.): TORVEASTRO 2023, MMS 130, pp. 27–34, 2023.
https://doi.org/10.1007/978-3-031-28447-2_2

The custody of ESCC specifications is the responsibility of ESA, which holds the copyright and publishes them through a program called the European Space Component Information Exchange System (ESCIES), whose activities are documented in [2, 3].

Another source of documentation is the site of the European Cooperation for Space Standardization (ECSS) initiative, also led by ESA, national space agencies and industry [4].

Table 1. Requirements and documents related to space station working environments

Document number	Document title	Notes
SSP 30243	Space Station Requirements for Electromagnetic Compatibility	MIL-E-6051D tailored for ISS
MIL-E-6051D	Electromagnetic Compatibility Requirements, Systems	
SSP 30245	Space Station Electrical Bonding Requirements	MIL-B-5087B tailored for ISS
MIL-B-5087B	Bonding, Electrical, and Lightning Protection, for Aerospace Systems	
SSP 30237	Space Station Electromagnetic Emission and Susceptibility Requirements	Derived from MIL-STD-461B
MIL-STD-461B	Electromagnetic Emissions and Susceptibility Requirements for the Control of Electromagnetic Interference	
SSP 30238	Space Station Electromagnetic Techniques	Derived from MIL-STD-462
MIL-STD-462	Electromagnetic Emissions and Susceptibility, Test Methods for	
SSP 30242	Space Station Cable/Wire Design and Control Requirements for Electromagnetic Compatibility	Unique ISS document similar to, but not derived from DOD-W-83575
DOD-W-83575	Wiring Harness, Space Vehicle, Design and Testing, General Specification for	
SSP 30240	Space Station Grounding Requirements	Grounding requirements unique to ISS

To formulate the design requirements of a spatial robot, an exhaustive investigation of these complex regulations would be necessary and other sources may be found in the documentation of other spatial agencies not listed above.

In order to describe the effect of the characteristics of such a special working environment on the design of the robot, we were consulting introductory documentation (see Table 1) and, where it was felt that further study was needed, the service robots in ISS equipment were considered, in particular the Canadarm2 (Figs. 1 and 2) arm and the Dextre (Fig. 3) robot of the Canadian Space Agency (CSA - ASC), the ERA robot (Fig. 4) of the European Space Agency (ESA), and the anthropomorphic robot Robonaut 2 (Fig. 5) of NASA.

Fig. 1. Canadian Space Agency astronaut David Saint-Jacques photographed Canadarm2, the robotic arm on the International Space Station, when it was fully extended. (17 m long) (Credit: Canadian Space Agency/NASA)

This paper thus describes the conclusions reached in this preliminary study for the working environment of the future robot and its launch to Low Earth Orbit (LEO) orbits, and then outlines the resulting requirements for the mechanical design of TORVEastro.

2 Characterization of the Low Earth Orbit (LEO) Working Environment

The International Space Station (ISS) (Fig. 6) occupies a LEO Earth orbit. Orbits of this type are characterized by an altitude between 160 and 1000 km [5]. The lower limit is imposed by the atmosphere, as excessive stress due to frictional forces would greatly reduce the lifetime of a station traveling in an orbit of that height, while the upper limit is imposed by the presence of the Van Allen belts, which would cause enough radiation exposure to impair the proper functioning of on-board equipment. Another advantage of this orbit is the limited propagation delay of radio transmissions (20–25 ms), comparable to that of some terrestrial links.

The ISS orbits about 400 km to 420 km above Earth's surface with an inclination of 51 degrees; to maintain itself on this orbit, the ISS must travel at about 27,400 km/h, with a revolution period of about 90 min.

Fig. 2. Astronaut Stephen K. Robinson, STS-114 mission specialist, anchored to a foot restraint on the International Space Station's Canadarm2 (*credit* Canadian Space Agency/NASA)

Fig. 3. Dextre, also known as the Special Purpose Dexterous Manipulator (SPDM), on the ISS (credit NASA)

The three main sources of charged particle radiation naturally occurring in space are galactic cosmic rays, solar proton events, and the trapped radiation belts. The ISS orbit is far from the inner Van Allen Belt (which begins around 1,000 km of altitude), but, without the protection of Earth's atmosphere, astronauts are exposed to about 1 mSv of radiation each day due to the constant stream of cosmic rays (a human being on Earth in a year receives an equivalent dose of 2.4 mSv from natural sources).

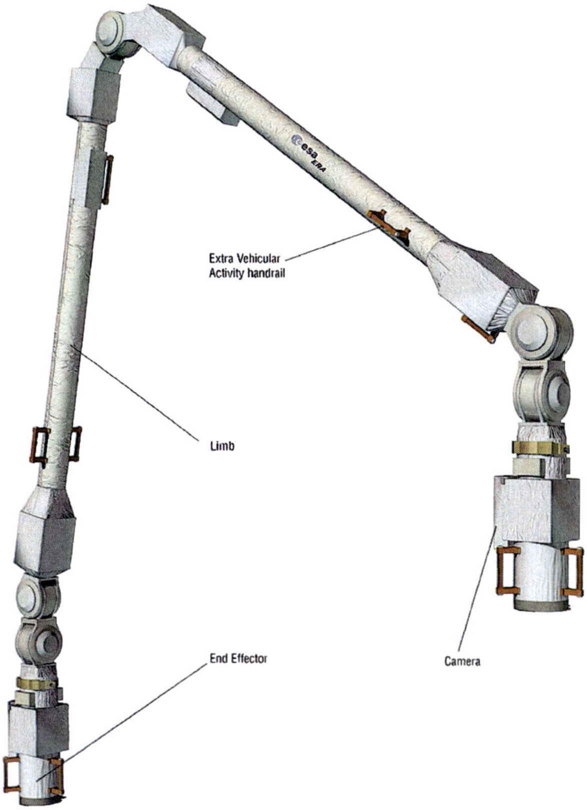

Fig. 4. ERA robotic arm layout (credit ESA)

Cosmic rays themselves damage electrical, electronic, and electro-mechanical components; the ESCIES program continuously updates the list of industrial components that have been qualified for use in space; for ionizing radiation compatibility the reference specification is ESCC Basic Specification No. 22900 (Issue 5) joined by MIL-STD-883, which is the standard for testing microelectronic devices in military and aerospace environments [6, 7].

ENEA joins the ASI Supported Irradiation Facilities [8] (ASIF) program, which was initiated as a cooperation agreement between Italian institutions involved in space and technology research and particle accelerators (ASI, ENEA and INFN); in particular, at the Casaccia Research Center, the Calliope facility is located, which constitutes a gamma irradiation facility (through Cobalt-60 sources) capable of performing qualification tests in accordance with ESA specifications by adopting for dose emitted in the unit of time the two standards: "standard dose rate" (36 Gy/h to 360 Gy/h) and "low dose rate" (0.36 Gy/h to 3.6 Gy/h) [9].

Out of ionizing radiation, another extreme characteristic of the environment around the ISS is the temperature, which can range from 100 °C (when the station is directly exposed to the sun) to −50 °C (when the station passes through the cone of shadow

Fig. 5. NASA/GM Robonaut 2 being upgraded on-orbit (credit NASA)

Fig. 6. The ISS pictured from the SpaceX Crew Dragon (credit NASA)

projected by the Earth); in addition, these two extremes alternate rapidly within the 90-min revolution period (35 of which are in the shade and the remainder in the sun).

The rapid temperature alternation results in continuous thermal expansion and contraction, which can cause first micro-fractures in the materials of construction and then real cracks under the action of dynamic loads. Thus, the importance of the choice of materials for the construction of service robots becomes evident.

By way of example, we point out that for the robotic arm Canadarm (supplied to NASA's Shuttles to unload loads transported to the ISS) titanium, stainless steel and carbon fiber reinforced epoxy resins (16 layers of resins and carbon fibers with high Young's modulus) were chosen as materials for construction. For its successor Canadarm2 [10] (mounted on the ISS), 19 layers of high-strength carbon fiber thermoplastics were used. Carbon fiber tubes with titanium fittings were also used for ERA (the ESA robot that reached the ISS recently).

In order to then ensure the proper functioning of the robot's electronics even in extremely cold conditions, it may be necessary to provide a coating of it with a heat-insulating material and use electric heaters controlled by thermostats. Again as an example, the Canadarm arm was entirely covered with a multilayer thermal insulation system consisting of alternating layers of gold Kapton, Dacron mesh fabric, and an outer covering of "Beta cloth" (a fireproof fiberglass fabric also used for some astronaut suits).

Another critical environmental aspect is due to the fact that the metal parts of objects in LEO orbits tend to accumulate electrical charges, both due to the interaction with the plasma generated by the action of ionizing radiation on the Oxygen and Nitrogen atoms (in turn produced by the action of UV rays on their diatomic molecules) and due to their fast motion in the Earth's magnetic field; therefore, the design will have to take this into account as well to avoid the possibility of discharge phenomena with electrical arcs between the robot and the space stations.

It will be preferable to operate with voltages of 28 V at most, a value traditionally used as the main voltage in the avionics industry and which has not generally resulted in observations of arcing phenomena except under extreme environmental conditions, characterized by magnetic perturbations, on spacecraft operating on orbits with high inclinations.

A final aspect that we would like to emphasize in this paper is the presence in orbit of space debris produced by old missions [11]; larger debris (above 10 cm) can be tracked and there are models [12] that, by continuously updating the calculation of their orbits, predict their arrival, thus allowing EVA activities to be planned safely. On the other hand, impact with smaller debris, characterized by more unstable orbits, is more difficult to predict and very dangerous considering that it can occur with very high relative velocities (about 10 km/s in low orbit; in practice, a 1-g debris possesses the kinetic energy of a 1000 kg car moving at 36 km/h).

3 Launch Characterization

To examine the design requirements of launching to a space station in LEO orbit, we can consider using the transport system to the ISS provided by the SPACE X company consisting of the FALCON 9 [13] launcher and the Dragon 2 spacecraft in Cargo Dragon (cargo-only transport) and Crew Dragon (mixed crew and cargo transport) versions [14].

During flight, the cargo carried by the launchers undergoes axial and lateral accelerations; the former are determined by vehicle thrust and drag, the latter by wind gusts, maneuvers, engine shutdown in the first phase of the launch, and other short-lived events.

For the Falcon 9 launcher, axial acceleration can range from −2 to 6 g and lateral acceleration can be up to 2 g, but in use with the Crew Dragon shuttle, the axial acceleration to which the astronauts are subjected (and thus also the load) is limited to 4.5 g.

Regarding vibration, in general, the loads carried by the Falcon 9, to avoid interaction with the launch vehicle dynamics, should maintain the primary lateral frequency above

10 Hz and the axial frequency above 25 Hz; the shock and vibration response at the time of separation of the two stages of the launcher for which SPACE X refers to the NASA-STD-7001 standard must also be considered.

4 Conclusion

The design of a new space station robot first prescribes a detailed analysis of the operating conditions under which the device will operate. In this paper, we have reviewed what is in the literature so as to circumscribe possible design and deployment errors.

Acknowledgments. The TORVEastro project activities reported in this paper were funded by Regione Lazio [Project 36517 POR FESR LAZIO 2014–20].

References

1. NASA Technical Standards Program (NTSP) website access page https://standards.nasa.gov/nasa-technical-standards. Accessed 28 January 2023
2. European Space Component Information Exchange System (ESCIES) website access page https://escies.org. Accessed 28 January 2023
3. European Space Components Coordination (ESCC) website access page https://spacecomponents.org. Accessed 28 January 2023
4. European Cooperation for Space Standardization (ECSS) website access page https://ecss.nl. Accessed 28 January 2023
5. Boldt-Christmas, L, LEO Earth orbit, ESA multimedia web page https://www.esa.int/ESA_Multimedia/Images/2020/03/Low_Earth_orbit. Accessed 28 January 2023
6. McCollum, M., Kim, L., Lowe, C.: Electromagnetic compatibility considerations for international space station payload developers. In: 2020 IEEE Aerospace Conference, pp. 1–9, IEEE, Big Sky, MT, USA (2020), https://doi.org/10.1109/AERO47225.2020.9172800. Accessed 28 January 2023
7. Calle, C.I.: The electrostatic environment at the international space station. In: Proceedings of the 2016 Electrostatics Joint Conference, Electrostatics Society of America, Purdue University, West Lafayette, IN, USA (2016), http://www.electrostatics.org/images/B1.pdf. Accessed 28 January 2023
8. ASI Supported Irradiation Facilities website access page http://www.asif.asi.it/. Accessed 28 January 2023
9. Baccaro, S., Cemmi, A., Di Sarcina, I., Ferrara, G.: Gamma irradiation Calliope facility at ENEA—Casaccia Research Centre, ENEA Technical Report (2019). https://www.enea.it/it/seguici/pubblicazioni/pdf-opuscoli/calliope.pdf. Accessed 28 January 2023
10. Canadarm, Canadarm2, and Canadarm3—A comparative table, ASC-CSA webpage https://www.asc-csa.gc.ca/eng/iss/canadarm2/canadarm-canadarm2-canadarm3-comparative-table.asp. Accessed 28 January 2023
11. About space debris, ESA webpage https://www.esa.int/Space_Safety/Space_Debris/About_space_debris. Accessed 28 January 2023
12. Johnson, N.L.: Space debris modeling at NASA. In: Proceedings of the 3rd European Conference on Space Debris, ESOC, Darmstadt, Germany, 19–21 March 2001, ESA (2001)
13. SpaceX, FALCON 9 webpage https://www.spacex.com/vehicles/falcon-9/. Accessed 28 January 2023
14. SpaceX, DRAGON webpage https://www.spacex.com/vehicles/dragon/. Accessed 28 January 2023

Dimensional Design of Torveastro, a Space Orbital Station Service Robot

Daniele Cafolla(✉) and Marco Ceccarelli

LARM2—Laboratory of Robot Mechatronics, University of Rome Tor Vergata, Rome, Italy
contact@danielecafolla.eu, marco.ceccarelli@uniroma2.it

Abstract. The development of space-intelligent robots for space exploration and utilization of space resources is essential for the advancement of space automation technology. As a result of design considerations, a kinematic design for a space orbital station Torveastro service robot is proposed, with a dimensional design suitable for use in space station service applications. Requirements and characteristics are reviewed to identify design issues and operational characteristics. A feasibility study is outlined using simulations to characterize the fundamental functioning and conduct a performance analysis.

Keywords: Space robotics · Service robotics · Robot design

1 Introduction

Space intelligent robots to achieve space exploration and space resource usage [1] is an essential approach for the development of space automation technology. In addition to the International Space Station (ISS), China successfully launched the Tianhe Core Module in 2021 and plans to construct a large-scale, long-term manned national space laboratory between 2021 and 2022 as it expands China's Space Station gradually (CSS). Currently, most ISS operations are performed by human astronauts. In a medium- to long-term microgravity environment, astronauts may experience a variety of physiological and pathological disorders that pose a grave threat to their health [2]. When performing duties outside of the cabin, however, they are exposed to significant health and safety dangers.

In addition, since the Soviet Union successfully launched the first satellite of humanity into space in 1957, humanity has launched about 7,000 satellites into space, of which around 10% have failed to orbit correctly or failed in the initial stage. Indeed, even with a high-reliability design, many spacecrafts fail in the hostile space environment, the majority of which are high-value, high-orbit satellites that are difficult to maintain, [3–5]. With the in-depth development of space resource consumption in the future, large-aperture antennas, large-diameter optical devices, large-scale solar power stations, and other large space structures will be important development goals of space management and conservation worldwide, while on-orbit assembly and maintenance of these systems will primarily rely on space intelligent robots. The development and use of space robots for on-orbit construction, assembly, and maintenance of space stations, satellites,

© The Author(s), under exclusive license to Springer Nature Switzerland AG 2023
M. Ceccarelli et al. (Eds.): TORVEASTRO 2023, MMS 130, pp. 35–44, 2023.
https://doi.org/10.1007/978-3-031-28447-2_3

and huge space structures is therefore of considerable social and economic importance, [6–8].

As shown in Fig. 1, an intelligent robot designed for outer space operations must meet several criteria [9–11]. Design requirements concentrate on the problem and the desired outcome, such as performing a movement from one location to another using existing structures such as handrails. Mechanism requirements focus on the creation of an optimal mechanism that can optimize the performance of the intelligent robot for the desired task, such as a compliant gripper that can adapt grasping while moving along handrails. The objective of electronics design is to create electronics that are compatible with the intended environment and that are protected from its characteristics, such as ionizing radiation. The ultimate objective of control design is to link everything together, such as receiving an input and executing the required task using embedded sensors. The hostile space environment, including microgravity, complicated illumination, and powerful radiation in space, presents obstacles for space robotics in large-range stable mobility, dexterity and safe manipulation, precision sensing, and high-precision measurement [10–13].

Hence, this paper presents a dimensional design of Torveastro, a space service intelligent robot for the on-orbit operation requirements of space stations, which is able to effectively utilize installations developed for human astronauts movement for outer space operations using handrails.

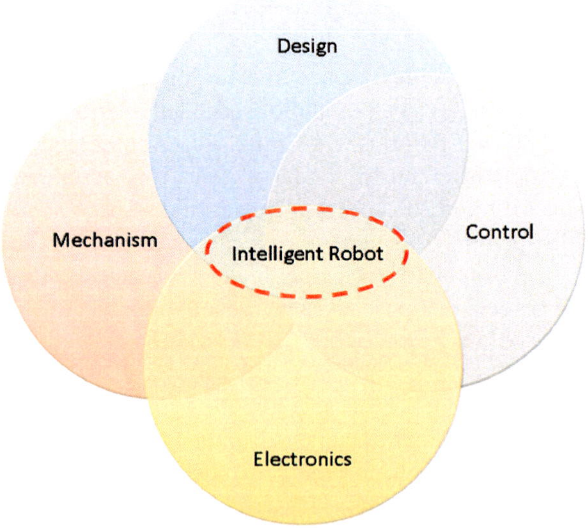

Fig. 1. A conceptual framework for outer space intelligent robot design.

2 Robotic System for Space Station Applications

Humanoid robot astronauts have been highlighted as one of the main pathways of development for intelligent space robots in major aerospace nations such as Europe, the

United States, and Russia. Since humanoid robots have a similar appearance to a human astronaut, there is no need to make any modifications for adapting to the space station. Controlling robots like astronauts to perform human-like movements and operations is an effective method for assisting or replacing astronauts on duty in the space station. Particularly, the NASA-developed humanoid robot astronaut Robonaut2 (R2) boarded the ISS in 2011 [3], where the technology of space intelligent operation is being tested. R2 is a mechatronics system with 42 independent DOF and over 350 sensors, comprising two arms with 7 DOF, two hands with 12 DOF, a neck with 3 DOF, and a waist with 1 DOF.

Additionally, the German Aerospace Center DLR has created a humanoid mobile robot named Justin [4]. It has a diameter of 2.70 m, a mass of approximately 200 kg, and a rated load of 20 kg. It has a total of 51° of freedom. Its bendable, lightweight arm and four-finger dexterous hand make it a perfect platform for studying manipulation dexterity. The motion detection sensor and stereo camera allow the robot to rebuild the environment in three dimensions.

The ASTROBOT robot astronaut created by Beijing Institute of Technology (BIT) weighs approximately 50 kg. It has two degrees of freedom for the head and six degrees of freedom for each arm. The master-slave arm's coordinated control algorithm enables the robot to execute a bionic climbing motion in a microgravity environment. To assure the stability of the climbing movement outside the cabin, a control system based on contact force-based feedforward compensation is implemented. In addition, a variety of image enhancement algorithms and human-like dexterous manipulation algorithms [5] have been developed to facilitate the installation and maintenance of extravehicular components under space-specific lighting conditions.

The robot astronaut exemplifies intelligent robotics for space station applications and is a research focal point for intelligent robots in space. However, it is challenging to perform a wide range of stable movements and safe dexterous manipulation in a limited microgravity space [6], and it is also challenging to check the 3D motion of the robot astronaut on the ground. However, the motion mechanism of humanoid robots in microgravity environments is vastly different from those on the earth. Concerning robot astronauts, forming a parameterized representation of the law, and studying the theory of motion planning and control appropriate to the space environment is an urgently needed scientific solution.

Currently, machine vision is an essential foundation for robotic manipulation, and the picture filtering, recognition, and precise positioning methods of the present machine vision [7] have made significant development. Unfortunately, there is a stark contrast between the shadow area and the illuminated area in space, and the mirror-like coating of the space device may cause multiple refraction and reflection of light, high-energy particle radiation, and other complex interference environments, which severely interfere with the characteristics of the operating objects and present a challenge for the existing image processing algorithms.

For instance, the R2 robot astronaut of the ISS is outfitted with four RGB cameras and an infrared TOF camera on its head to offer accurate depth information for enhancing measurement accuracy. However, R2 struggles to overcome the interference of the complicated spatial environment.

With the advancement of intelligent robot technology, the space station's intelligent robot has improved in terms of stable motion, dexterous manipulation, and accurate sensing and measuring. The purpose of this paper is to provide a dimensional design for a robot that can move and operate along existing handrails built outside the space station for astronaut movement. This was done by combining bionics and robotics to investigate the theories and methods of human-like stable motion.

3 Dimensional Design

Since the handrails designed for human movements in outer space are 1 m [8] distant between each other, the total distance between the astronaut service robot's extended effectors must be 1 m when a minimum diameter is considered. Figure 2 shows the handrails installed in outer space.

It has been decided that the body of the service robot would be cylindrical and consist of three limbs that are 120° apart and equally spaced. In addition, the arm will have two links and an end effector. The handrails are installed along the surface of the station to allow astronauts to move in outer space to perform maintenance and service operation. The handrails have a compact structure of (34.95 × 15.9) each and are very stable and easy to grasp.

Figure 3 shows a scheme for dimensional design where the arc-shaped links are conceived for a compact home configuration design that is suitable for launching storage.

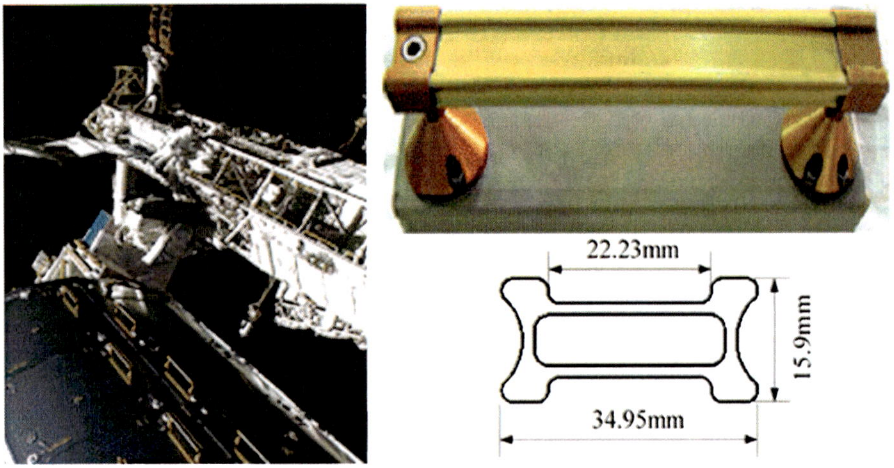

Fig. 2. Examples of handrails for astronaut robot motion detail, [8].

By using information in Fig. 3, a geometric optimization according to Eqs. 1–3 can be performed to determine the robot body and link dimensions:

$$c = 2(R + d)\sin\alpha \tag{1}$$

$$l' = c \tag{2}$$

$$l'^2 + L^2 - 2l'L\cos\theta = (R + f)^2 \tag{3}$$

where

- c is length of the entire arm
- R is the radius of the body
- d, f are the distance from the connection of the arm to the body
- α is the rotation of the body in respect to the horizontal reference frame
- $2L$ is the distance between the handrails
- l' is the sum of the cord of the arm link
- θ is the inclination of the cord of the arm link in respect to the horizontal reference frame.

The horizontal reference frame indicates the distance between the handrails, with the beginning and ending points indicating the grasping points. The configuration with the body rotated by 120.00° and the arm extended and slanted by 25.62° was chosen because it reflects the worst-case scenario in which the robot is moving between two known places using handrails.

In addition, L that is 500 mm, half of the distance between the handrails, and that d and f have the same length of 40 mm.

Imposing that $l' = c$, after solving Eq. 1 using the Carnot's Theorem presented in Eq. 3 it is possible to compute the optimal radius of the body R and length of each link of the arm $c/2$. The results give an optimal radius of the body R = 184.78 mm and a length of the link of the arm $c/2 = 194.67$ mm. By using these parameters, a kinematic simulation has been carried out to compute the workspace of the proposed mechanism and its angular velocities and angular accelerations to check the design feasibility and to characterize the operation performance.

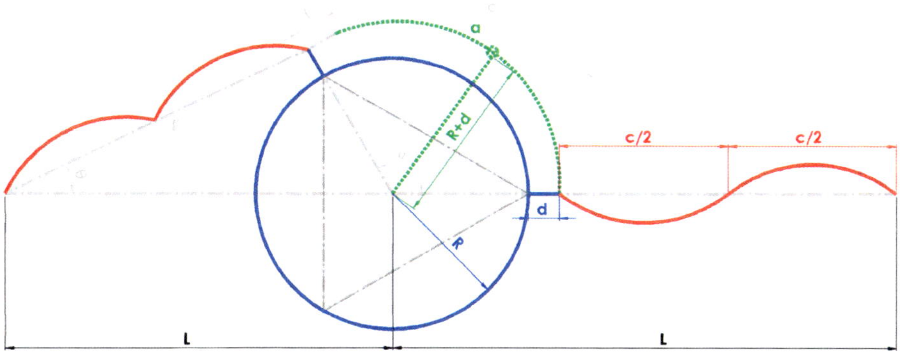

Fig. 3. Scheme for dimensional design.

4 A Kinematic Simulation

The design values derived from the above conceptual scheme have been approximated to yield a body diameter of 370 mm and a link length of 195 mm, resulting in a total arm length of 390 mm. All the mentioned parameters have been used to design a 2D CAD model to perform a kinematic simulation, as reported in Fig. 4.

The simulation illustrates only one arm during a 10-s closing operation that returns the arms to a launch configuration in which the arms are fully wrapped-around the body for the most straightforward operations. The movement starts with the arms totally opened in Fig. 4a. In the second step of the movements the arms are moved at the same speed, the arm finishes its movement after moving by 70 deg arriving at the middle position in Fig. 4b. In the approaching position in Fig. 4c, the arm remains standstill while the forearms continue to move up by an additional 14 deg to the contact with the robot body. Finally in the closing position in Fig. 4d only the end-effector moves by 88° to arrive at the full-closed launching configuration.

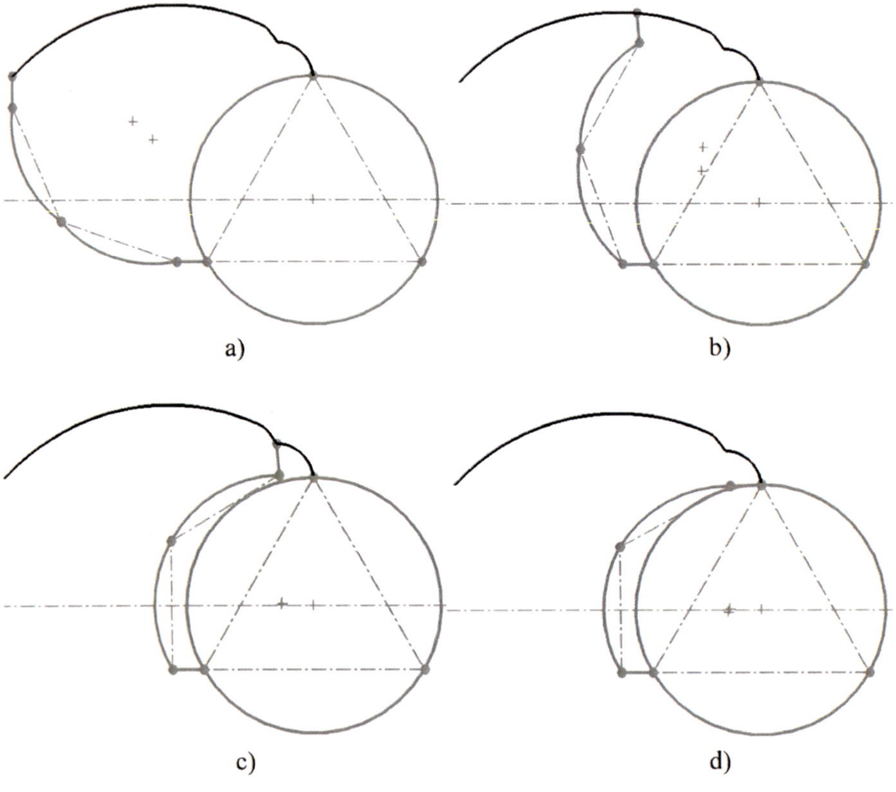

Fig. 4. Kinematic simulation snapshots: (a) starting position, (b) middle position, (c) approaching position, (d) closed position.

Figure 5 illustrates the angular velocities of the arm's three segments: the arm, the forearm, and the end-effector (EE). During the closing procedure, it is important to note that the arm moves from second 0.00 to second 3.00, reaching a maximum of −35.00 deg/s. The forearm rotates continuously from second 0 to second ten, achieving a maximum of −2.10 deg/s. In the last phase of the closing operation, the EE is activated at second 8.00 up to second 10.00 with a maximum velocity of −43.93 deg/s.

Figure 6 shows the angular accelerations of the arm's three segments. The movement of the arm presents one step of angular acceleration from second 0.00 to second 1.50 reaching a maximum value of −45.75 deg/s^2; the second step from second 1.50 to second 3.00 represents the deceleration reaching a maximum value of 46.36 deg/s^{2}until stopping allowing the forearm to proceed for a more precise closure avoiding impacts to the body. The linear angular acceleration reflects the constant and smooth behavior represented by the angular velocities.

The forearm as shown in the velocity trend is moving constantly from second 0 to second 10, this is reflected in a linear angular acceleration from second 0 to second 5 and a linear deceleration from second 5 to second 10 ranging from −0.83 deg/s^2 to 0.83 deg/s^2. Finally, the EE while actuated presents an angular acceleration from second 8.00 up to second 10.00 reaching a maximum value of 57.00 deg/s^2.

The computed kinematic simulation demonstrated that the intended operation had angular velocities and angular acceleration that are achievable with commercially available components.

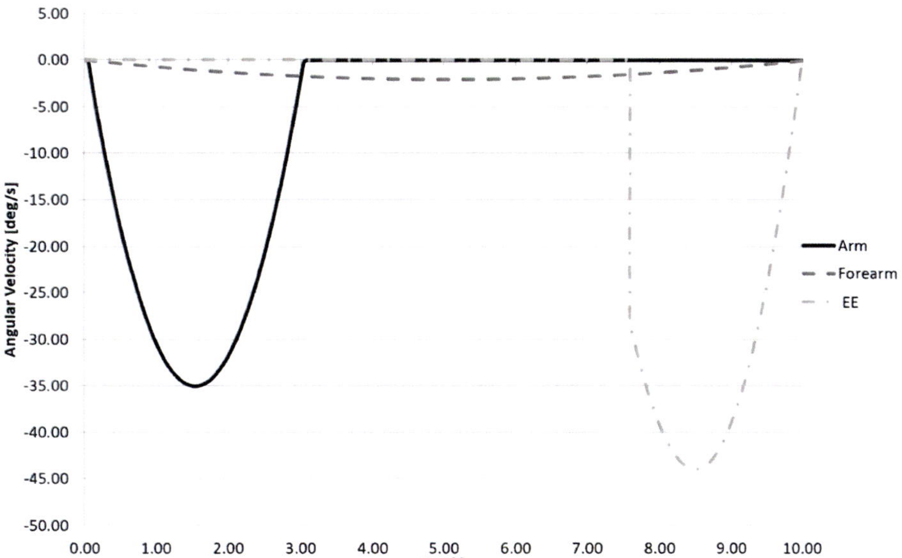

Fig. 5. Computed results for the kinematic simulation in Fig. 3 for angular velocities.

In addition, a layout in Fig. 7 is proposed dimensioning and positioning three motor units acting the 3° of Freedom (DoFs) per arm, the control unit, the vision system, and the battery. All electronics, actuators, and controllers will be protected and housed within the body. The layout is established with consideration for the size of the three commercial actuators Dynamixel MX-64T that will be housed within the motor unit. Additionally,

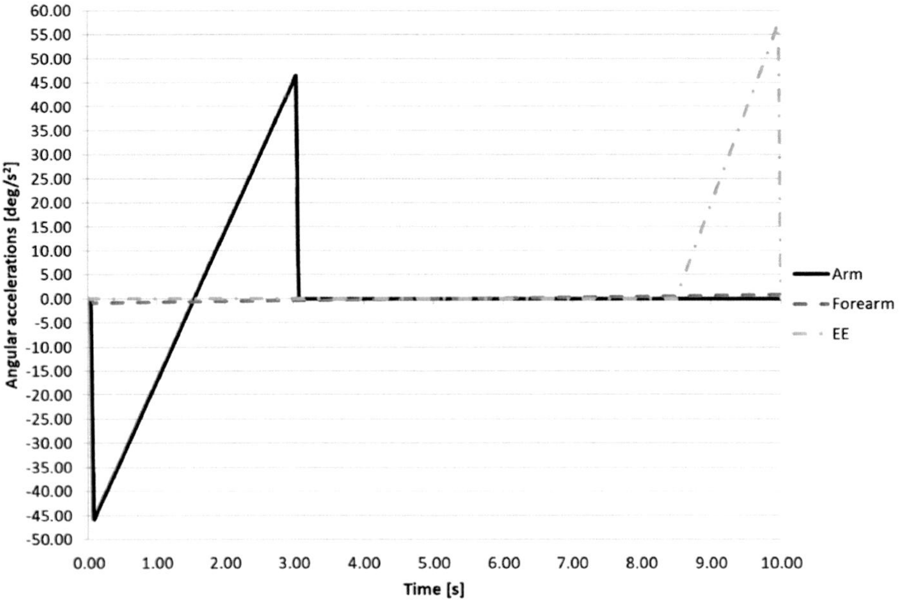

Fig. 6. Computed results for the kinematic simulation in Fig. 3 for angular accelerations.

the size and placement of three LiPo 14.8V 6200mAh batteries that will be used for optimal operational autonomy have been considered. A Raspberry Pi 4 will be used as a microcontroller and connected to three Raspberry cameras and will be placed in the middle of the body. The designed prototype weighs approximately 50N. Summarizing, in consideration of the calculated results, commercial components have been selected to manufacture a lab prototype for testing in terrestrial gravity field. The characteristics are shown in Table 1.

Table 1. Main prototype characteristics of the lab prototype.

Description	Value
Body diameter	ϕ370 mm
Body height	200 mm
Body mass	$\approx$5 kg
Microcontroller	Raspberry Pi 4
Webcams	Raspberry Pi camera
Battery	LiPo 14.8V 6200 mAh

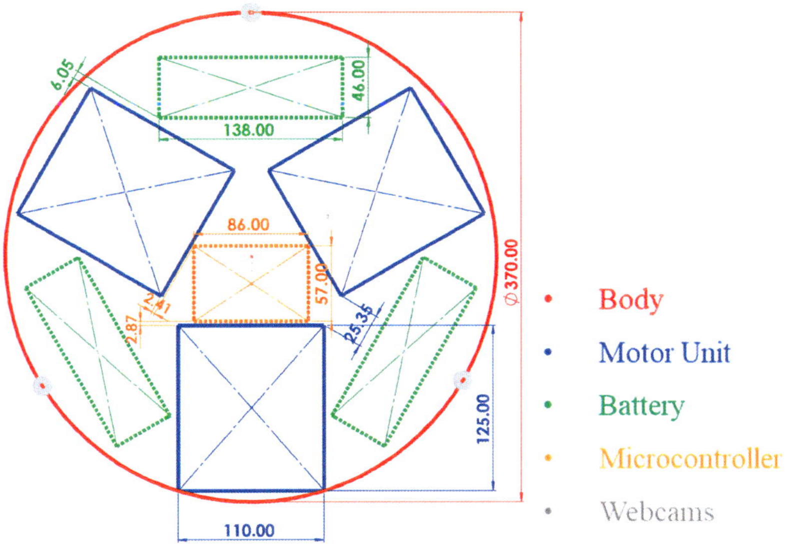

Fig. 7. A design scheme of the Torveastro body with inside component location.

5 Conclusions

A dimensional design of Torveastro a space orbital station service robot is presented, along with the computation and modeling of its design parameters. Design difficulties and solutions are explored to highlight the dimensional design computing parameter necessary to produce a lab prototype. Kinematic simulation results indicate that the proposed method is suitable in terms of motion performance with angular velocities and angular accelerations that are feasible for commercial components in space-like operations. Future efforts will rely on CAD design and dynamic simulation, as well as the fabrication and testing of the complete 3D prototype.

References

1. Bluethmann, W., et al.: Robonaut: a robot designed to work with humans in space. Auton. Robot. **14**(2), 179–197 (2003)
2. Tan, C., et al.: Investigation on effects of gradually increased Coriolis stimulus vestibular rehabilitation training. Hangtian Yixue yu Yixue Gongcheng **26**(1), 20–23 (2013)
3. Diftler, M.A., Mehling, J.S., Abdallah, M.E., Radford, N.A., Bridgwater, L.B., Sanders, A.M., Askew, R.S., Linn, D.M., Yamokoski, J.D., Permenter, F.A., Hargrave, B.K.: The robonaut 2 hand-designed to do work with tools. In: 2011 IEEE International Conference on Robotics and Automation, pp. 3425–3430. IEEE (2011)
4. Borst, C., Wimbock, T., Schmidt, F., Fuchs, M., Brunner, B., Zacharias, F., Giordano, P.R., Konietschke, R., Sepp, W., Fuchs, S., Rink, C.: Rollin'justin-mobile platform with variable base. In: 2009 IEEE International Conference on Robotics and Automation, pp. 1597–1598. IEEE (2009)
5. Wei, B., Jiang, Z., Li, H., Dong, Q., Ni, W., Huang, Q.: Adaptive impedance controller for a robot astronaut to climb stably in a space station. Int. J. Adv. Rob. Syst. **13**(3), 81 (2016)

6. Flores-Abad, A., Ma, O., Pham, K., Ulrich, S.: A review of space robotics technologies for on-orbit servicing. Prog. Aerosp. Sci. **68**, 1–26 (2014)
7. Lowry, S., et al.: Visual place recognition: a survey. IEEE Trans. Rob. **32**(1), 1–19 (2015)
8. Ceccarelli, M., Li, H., Carbone, G., Huang, Q.: Conceptual kinematic design and performance evaluation of a chameleon-like service robot for space stations. Int. J. Adv. Robot. Syst. **12,** 17 (2015)
9. Araque-Isidro, J.E., Cafolla, D., Ceccarelli, M.: Problems and requirements for outer space astronaut service robot. In: Advances in Italian Mechanism Science. IFToMM Italy 2022. Mechanisms and Machine Science, vol 122. Springer, Cham (2022)
10. Hirzinger, G., Sporer, N., Schedl, M., Butterfass, J., Grebenstein, M.: Robotics and mechatronics in aerospace. In: 7th International Workshop on Advanced Motion Control. Proceedings (Cat. No. 02TH8623), pp. 19–27 (2002)
11. Zhang, T., Chen, Z., Wang, X., Liang, B.: Overview and prospect of key technologies of teleoperation of space robot. Aerosp. Control Appl **40**, 1–9 (2014)
12. Ceccarelli, M., Li, H., Carbone, G., Huang, Q.: Conceptual kinematic design and performance evaluation of a chameleon-like service robot for space stations. Int. J. Adv. Robot. Syst. **12** (2015)
13. Musabayev, T., Zhantayev, Z., Grichshenko, V.: Complex influence of space environment on materials and electronic devices in the conditions of microgravity. Adv. Space Res. **58**(6), 1138–1145 (2016)

Assembly and Testing of a New Joint Mechanism for TORVEASTRO

Jorge E. Araque-Isidro(✉) and Marco Ceccarelli

LARM2—Laboratory of Robot Mechatronics, University of Rome Tor Vergata, 00133 Rome, Italy

jorgeenrique.araqueisidro@students.uniroma2.eu, marco.ceccarelli@uniroma2.it

Abstract. This paper presents two joint mechanisms driven by cables that are designed for a cable-driven revolute joint. The proposed solution is analysed to characterize its operation referring to problems and requirements for its implementation in the limbs of the TORVEASTRO astronaut robot. The mechanical design is verified through a dynamic simulation whose results give a characterization of the operation performance.

Keywords: Space robots · Design · Joint mechanism · Prototype

1 Introduction

Mechanism actuated by cables has been adopted in many robotic manipulation systems for example [1]. These kinds of mechanisms also referred to as tendon-driven or wire-driven mechanisms, are usually driven by cables and have applications in a wide range of robot structures, as outlined in [2]. Research on the mechanism actuated by cables has been developed in the past 30 years and theoretical studies have been published on them, as reported for example in [2–12].

In particular, those mechanisms are used to work in harsh environments. Some harsh environments include nuclear power plants, toxic chemical industries, aircraft parts assembly, rescue, and space exploration [3]. In addition, cable actuation requires antagonism operation.

The joint with antagonist actuator concept is biologically inspired by the musculoskeletal system of animals [4], the actuators are placed at the base, and the force is transmitted through cables [1]. Using this motion and force transmission is necessary to adopt antagonism motion on the mechanism [4, 5].

Furthermore, when the cables take the place of the rigid links, the manipulators have much smaller inertia and a higher payload-to-weight ratio, which provides high speed and acceleration of the end-effector as outlined in [6, 7, 14].

In this paper, a new joint mechanism is proposed for motion and force transmission in manipulator arms where it is necessary to have actuated joints as in the limb design of the TORVEASTRO astronaut robot [13]. The proposed joint mechanism includes two antagonism movements by two different sliders-crank linkages that are used to generate

© The Author(s), under exclusive license to Springer Nature Switzerland AG 2023
M. Ceccarelli et al. (Eds.): TORVEASTRO 2023, MMS 130, pp. 45–59, 2023.
https://doi.org/10.1007/978-3-031-28447-2_4

rotation on a mobile platform as presented in [13]. The conceptual approach, the CAD design, and the performance analysis of the joint mechanism actuated by cables are discussed with results to show the design's feasibility and efficiency.

2 Requirements

Design issues are directly related to the operating features of a robotic limb. The main goal considered for this design is that the joint mechanism should be compact. A compact design can be considered by sizing the components so that the overall system has limited volume and at the same time, a proper motion range in terms of angular displacement as needed to be useful for a robotic arm.

Furthermore, motion transmission is required for the design of an actuated joint where specific motion is needed for a specific task in a robotic arm in terms of angular displacement. The cable-driven can be a solution when the actuator is attached at the base of the robot so that unnecessary load in the joint is avoided.

In general, a joint mechanism should be an efficient system that can manage a revolute joint expecting less possible motion and torque transmission. A summary of the requirements for a joint mechanism is shown in Fig. 1.

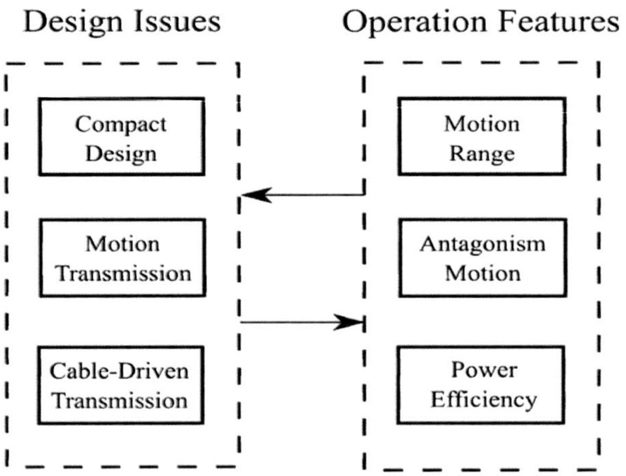

Fig. 1. Requirements for the design of a joint mechanism.

Referring to Fig. 1 operation features can be expected in a proper motion range of the limb through its joints while spanning in a proper workspace area for the planned locomotion and manipulation tasks. Considering the cable actuation, the efficient operation can be designed with a cable system whose function is based on the pulling actions of antagonist cables to keep the overall design still compact and lightweight. Main attention as per aerospace system will be devoted to low levels of power consumption, due to the limit of available power in an orbital station.

3 Design Solutions

Two solutions are investigated with conceptual design to evaluate the main characteristics of a final design solution.

The design of the joint mechanism is carried out considering the need to transmit force and movement from a servomotor to a joint platform through the antagonistic movement in two prismatic joints. A motor is used to transmit the movement and the necessary force, which allows the final output movement on the platform. The angular displacement of the motor will be converted to linear movements that allow the required output.

3.1 First Design

The first solution is designed with the conceptual design in Fig. 2.

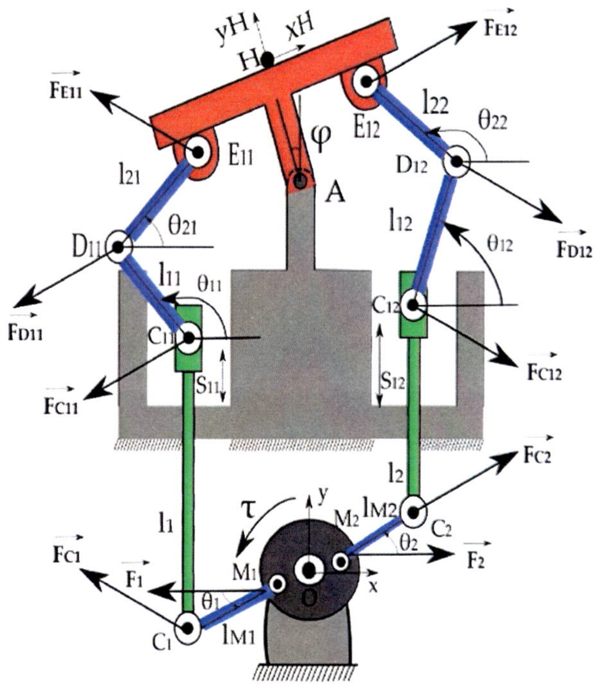

Fig. 2. Kinematic scheme of the first proposed joint mechanism.

The axis of the servomotor generates a torque, τ, a revolute motion. This revolute movement is transmitted through the pistons guides in C_{11} and C_{12} that perform a linear motion through the displacements S_{11y} and S_{12y}. This allows the transmission of the revolute movement into a linear one of a prismatic joint within the body frame of the joint mechanism system. The servomotor is connected at two links with revolute joints, which, when the revolute movement is generated, give in performing antagonistic linear movements.

The prismatic guides are connected to links through a revolute joint, which is the connection between the C_{11} body with l_{11} and l_{21} links and the C_{12} body with l_{12} and l_{22} links. The links l_{11} and l_{12} are of different sizes, to generate angles between the links that help the movement and force to be transmitted in a better way. But the links l_{11} and l_{21} as well as links l_{21} and l_{22} are of the same sizes, as are links l_{12} and l_{22}, to have the two mechanisms equal and running similarly but in antagonism.

The angle θ_{11} is the angle between body D_{11} and link l_{11} and in the same way, the angle θ_{12} is between body D_{12} and link l_{21}. Angles θ_{21} and θ_{22} are between links l_{11} and l_{21} and between links l_{21} and l_{22}.

The frame body of the joint mechanism system is connected to the mobile platform in the revolute joint point A. At the end of the platform E_{11} and E_{12} are the connection points of the two mechanisms working in parallel. Therefore, when the revolute movement is generated in the actuator disk, the mobile platform rotates with the angular displacement described by the angle φ.

3.2 Second Design

The second solution is designed with the conceptual design in Fig. 3.

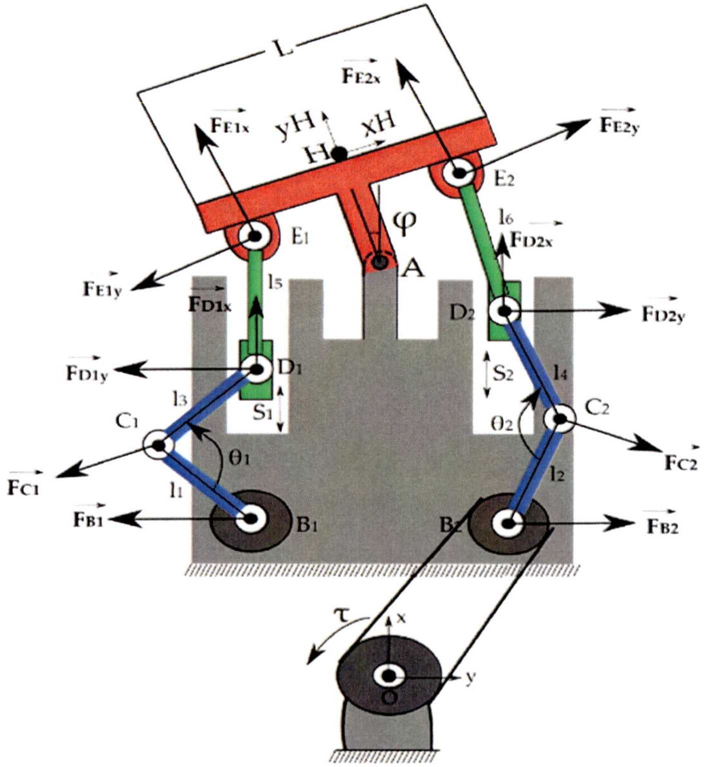

Fig. 3. Kinematic scheme of the second proposed joint mechanism.

The second solution's main difference in Fig. 3 concerning the above first design is that the joint mechanism is designed with component linkage in a different order, as explained below.

The kinematic design of the proposed second joint mechanism is shown in Fig. 3 with its parameters. The input torque is indicated by τ and is given by a servo motor on which a reference frame O_{xy} is assumed. The servomotor transmits the motion by a pulley that is connected to another two pulleys at points B_1 and B_2. This actuation makes the mechanism cable actuated. The motion is transmitted to the links l_1 and l_2 of the same length through the reaction forces $\mathbf{F_{B1}}$ and $\mathbf{F_{B2}}$ at the joints B_1 and B_2. The angle θ_1 is between the l_1 and l_3 links; similarly, the angle θ_2 is between the l_2 and l_4. The revolute joints C_1 and C_2 connect the links l_1 and l_2 with links l_3 and l_4. At the C_1 and C_2 joints act reaction forces, $\mathbf{F_{C1}}$ and $\mathbf{F_{C2}}$. The revolute joints in D_1 and D_2 connect the links l_3 and l_4 with links l_5 and l_6. On the D_1 and D_2 reaction forces, $\mathbf{F_{D1}}$ and $\mathbf{F_{D2}}$ move links l_5 and l_6 with an antagonist motion.

The motion of prismatic joints with points D_1 and D_2 is represented by the strokes S_1 and S_2. $\mathbf{F_{D1}}$ and $\mathbf{F_{D2}}$ are the reaction forces at the revolute joints in D_1 and D_2. The revolute joints at D_1 and D_2 connect l_3 with link l_5 and l_4 with link l_6 respectively. The links that connect the prismatic joint in D_1 and C_2 with the mobile platform with H reference points are given by l_5 and l_6, with the same length. The link l_5 is connected to the joint platform by the revolute joints E1 and similarly, l_6 by point E_2. $\mathbf{F_{E1}}$ and $\mathbf{F_{E2}}$ are the reaction forces at the revolute joints in E_1 and E_2 giving the motion to the mobile platform. The joint at point A is the mechanism-assisted revolute joint connecting the mobile platform to fixed housing with a relative revolute motion described by angle φ.

4 CAD Designs

Once the kinematic structure is designed with the required characteristics as in Figs. 2 and 3, a CAD model has been elaborated to be analysed kinematically and the dynamic mechanism is to check the feasibility and performance of the chosen final solution of Fig. 4. CAD design of the first solution is in Fig. 5.

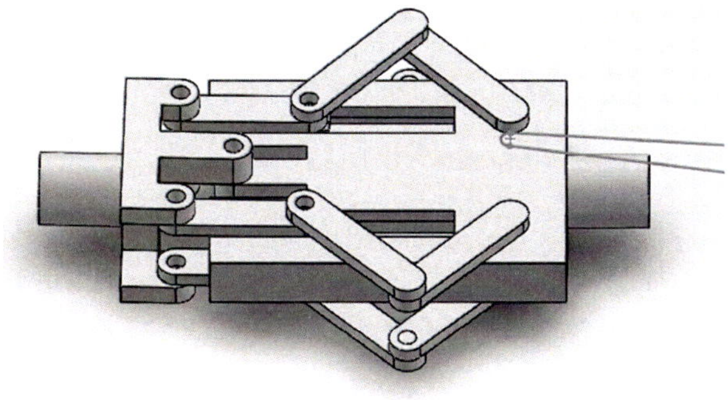

Fig. 4. CAD design of the chosen final solution from Fig. 3.

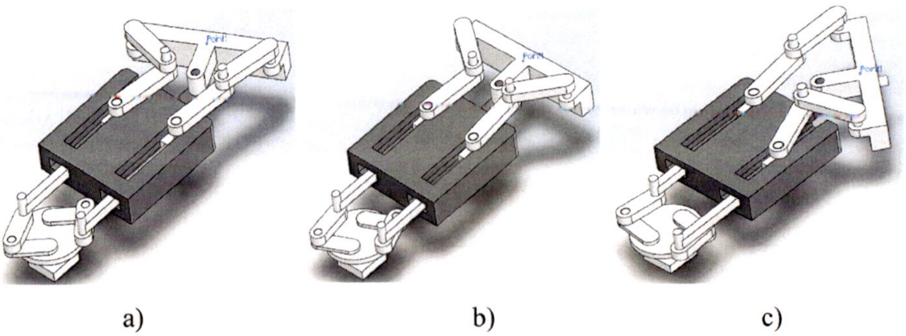

Fig. 5. A snapshot of the simulated operation of the first solution in Fig. 2 with the second operation mode in Fig. 6, as a function of the input angle: (a) −45°; (b) 0°; (c) 15°; (d) 45°.

The kinematic simulation of the joint mechanism is carried out using four operation modes, as summarized in Fig. 6. In the first operation mode, the links on both sides are in angles $\theta_1 < 180°$. y $\theta_2 < 180°$; in the second operation mode the links on both sides are in angles $\theta_1 > 180°$ y $\theta_2 > 180°$; In the third and fourth operation modes the links are in $\theta_1 < 180°$ y $\theta_2 > 180°$ and $\theta_1 > 180°$ y $\theta_2 < 180°$.

By simulating these different modes of joint mechanism operation, it can be detected that in the third and fourth modes, the links approached or passed through a point of singularity. In this scenario, the point of singularity occurs when the links l_1; l_3 and l_2; l_4 are aligned at an angle of 180°. With each other. Thus, the kinematic simulation showed that the first and second modes can perform the required movements without approaching the singularity point.

The mobile platform is connected to the frame body and the slider-crank mechanisms at three different points E_{11}, E_{12}, and A, respectively. The revolute joint in A provides the circular output movement and the slider-crank mechanisms provide the force to move the mobile platform. Therefore, the slider-crank mechanisms must work in an antagonist way.

The characteristic operation is analysed by choosing the first operation mode with a simulation as in Fig. 7, where the adopted joint mechanism is designed with the parameters listed in Table 1 to fit constraints in operation range and space availability for a mechanical design according to the issues in Fig. 1. The operation workspace range of the chosen joint mechanism design is computed in the simulation within 11 x 15 mm.

5 Performance Analysis via Simulation

A multibody dynamic simulation of the two designs was performed in the SolidWorks environment to characterize the motion performance with the data in Table 2 to determine the range of motion that the joint mechanisms can achieve in their performance. The simulation was carried out considering the analysis made with the previously chosen mode of operation for both joint mechanism solutions in Figs. 2 and 3 to characterize both solutions towards a selection for a final design, as reported in Fig. 4. CAD design of the first solution is in Fig. 7 and the CAD design of the second one as final solution

Table 1. Dimensions of the joint mechanism solution in Figs. 3 and 4.

Parameter	Value (mm)
B2	10
l1 = l2 = l3 = l4	20
l5 = l6	17.5
S1 = S2	16.42
L	20
A H	15

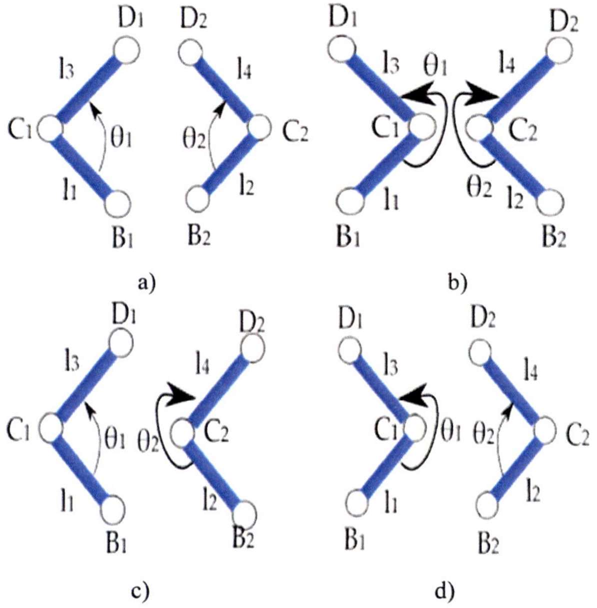

Fig. 6. Operation modes of the joint mechanism in Fig. 2 and Fig. 3: (a) $\theta_1 < 180°$ y $\theta_2 < 180°$; (b) $\theta_1 > 180°$ y $\theta_2 > 180°$; (c) $\theta_1 < 18°$ y $\theta_2 > 180°$; d) $\theta_1 > 180°$ y $\theta_2 < 180°$.

is in Fig. 4. The simulation models were completed with material selection for the joint mechanism components as PLA which is chosen to build a lab prototype by 3D printing manufacturing. The material data characteristics for the simulation are listed in Table 2.

The operation analysis of the first joint mechanism design is completed with a dynamic simulation to obtain results in terms of force transmission too. The main results of the simulation are reported in Figs. 8, 9, and 10.

The input for this simulation was the angular displacement of the motor, which was done for 5 s and went from −45° to 0°, for the first joint mechanism solution simulation results are reported in Fig. 8. The input motor angular displacement is shown in Fig. 8a and the required torque was computed as in the plot result in Fig. 8b. The forces transmitted by the actuator to the joint mechanism are analyzed in the reaction of

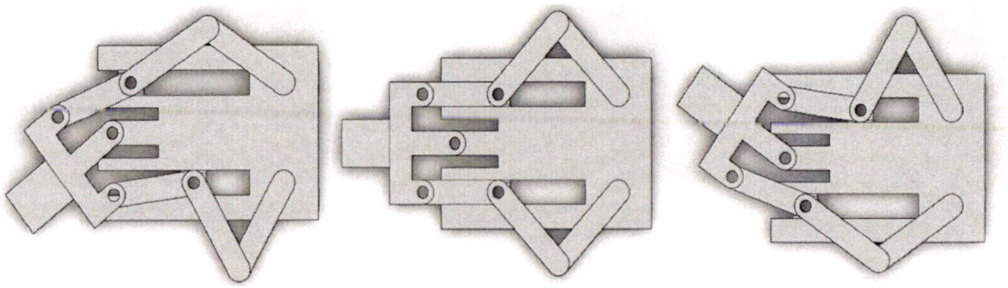

Fig. 7. A snapshot of the simulated operation of the design in Fig. 4 with the first operation mode in Fig. 6, as a function of the input angle: (a) 15°; (b) 0°; (c) −15°.

Table 2. Parameters of the joint mechanism material.

Parameter	Value
Dynamic friction coefficient	0.1
Static friction coefficient	0.15
Elastic modulus	2000 N/mm^2
Poisson's ratio	0.394 N/A
Shear modulus	318.9 N/mm^2
Mass density	1020 kg/m^2

the force that occurs at C_{11} and T_1, which is $\mathbf{F_1}$ is shown in Fig. 9. Likewise, the reaction of the force $\mathbf{F_2}$ that occurs in C_{12} and T_2, is shown in Fig. 10.

The forces $\mathbf{F_1}$ and $\mathbf{F_2}$ are analysed according to the plots obtained in Figs. 9 and 10. One can note that the force transmission ratio of 1:1. Force decreases occur at some points, for example at 1.5 s an input T_1 of 3 N of tension is presented and 1.4 N is transmitted to Force $\mathbf{F_1}$, and at the second 3 there is an input of T_1 of 2.8 N of tension and 1.4 N is transmitted to the force $\mathbf{F_1}$ in Fig. 13. Between 2.0 and 2.5 s, the input T_2 is 1.5 N, and the transmitted force is almost 0 N in Fig. 9. This is because the angles θ_{21} y θ_{22} of the links l_{11} l_{21} and l_{12}-l_{22} present rapid changes and thus the tension that maintains the transmission is lost.

The chosen joint mechanism gives input through the motion of the actuating cable from the pulley on the motion axle with a time operation in Fig. 11a between ±15°. The required torque was computed from the simulation as in Fig. 11b with a maximum value of 2029.86 N mm.

The angular displacement φ in the mobile platform is computed from 43° to −43° as shown in Fig. 12 as a convenient rotation for the joint platform. The linear displacement of strokes S_1 and S_2 are obtained as the expected antagonism motion as in Fig. 13 with a motion range of 16.42 mm.

The reaction force $\mathbf{F_{D2}}$ in its x component is transmitted by link l_6 throughout the simulation with the same value to the revolute joint E_2 whose reaction force $\mathbf{F_{E2}}$ is in

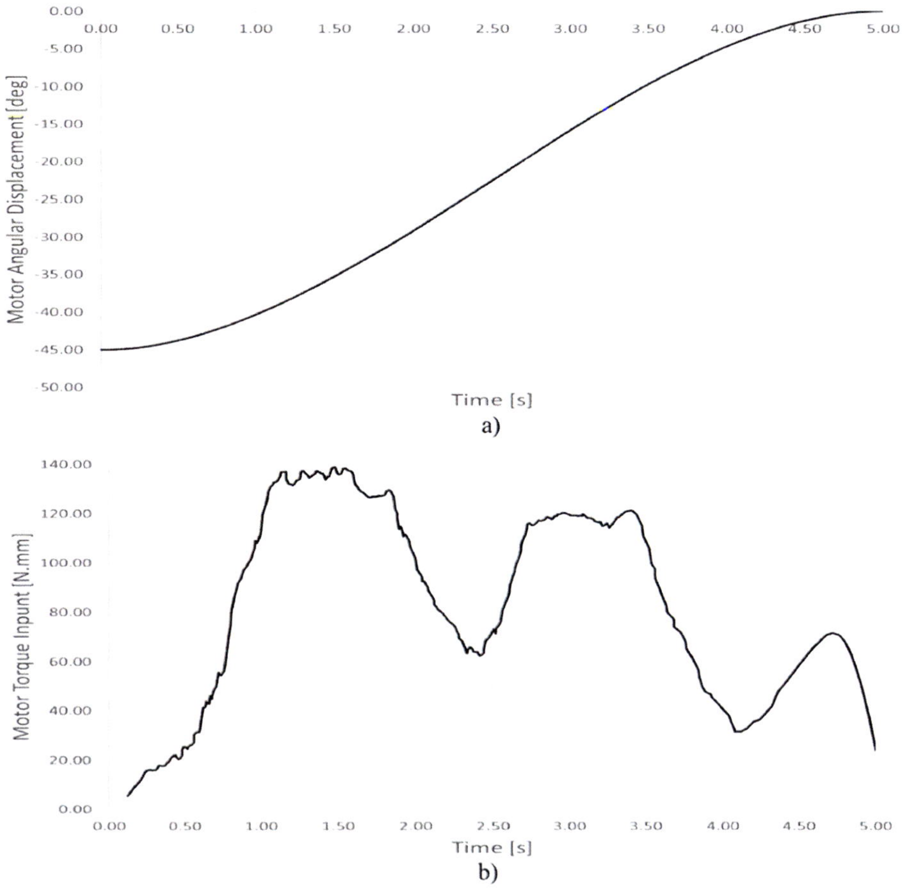

Fig. 8. Data of the driving motor for simulated operation of the joint mechanism solution in Fig. 2: (a) Input angle; (b) Computed torque during simulation in Fig. 5.

its x component as shown in Fig. 14a. On the other hand, the reaction force $\mathbf{F_{D2}}$ in its y component decreases as in the computed results, although it is small in the first stage from 0 to 2.5 s. in the second stage from 2.5 to 5.0 s the reaction force is transmitted with a significant decrease of t the reaction force $\mathbf{F_{E2}}$ in its y component.

6 Prototype

Prototypes of the joint mechanisms are built using 3D printing as a testbed in the laboratory for testing and acquiring data for performance analysis. The material used in these testbeds was PLA, its features are listed in Table 2. This material is used to build these prototypes because the TORVEASTRO astronaut robot in all its structural components is built of this material too.

Prototypes of the two solutions are built as in Figs. 15 and 16 as to be used in the validation of functioning and later experimental characterization via testing.

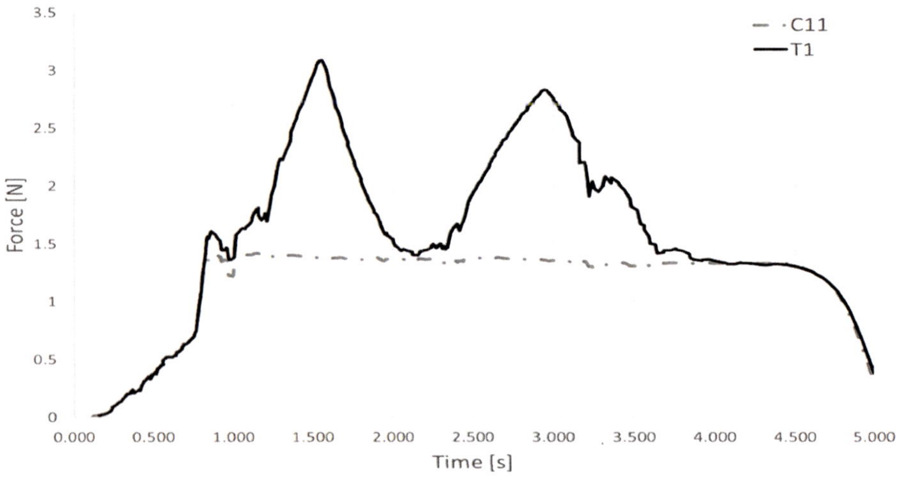

Fig. 9. Computed F_1 on the slider C_{11} and cable tension T_1 in Fig. 2.

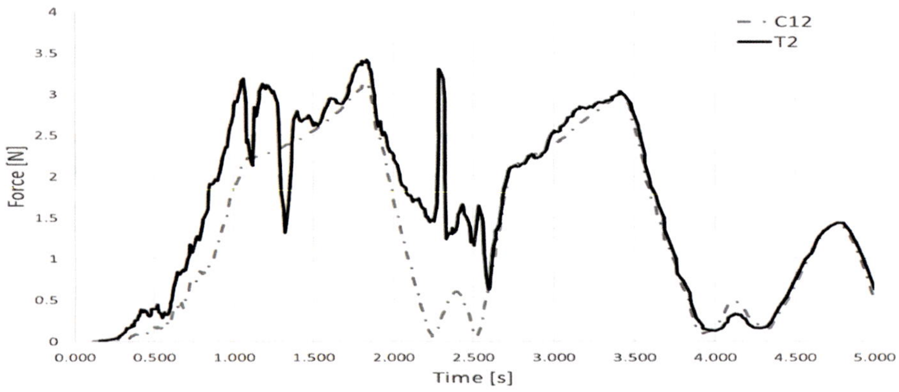

Fig. 10. Computed F_2 on the slider C_{12} and cable tension T_2 in Fig. 2.

Figure 15 shows the first prototype that was built at LARM2 as a validation of the designed solution. The solution was tested with a structure based on a direct slider-crank mechanism with configurations that avoid singularity problems and allow a compact mechanical design. In this testbed, a servomotor was selected to transmit the motion through the links simulating cables. The servomotor is controlled by an Arduino nano. The area of the first joint mechanism design is 120 × 100 mm. The maximum torque of the actuator is 1.5 Nm when the voltage of the servomotor is 12 V and the current is 1.5 A. when the prototype has been assembled some issues related to the proper motion of the antagonism mechanisms were discovered. The friction between the material affects the motion and force transmission. The friction features can be seen in Table 2. Another issue with the first design is that the joints connected to the mobile platform can have singularity points when the mechanism starts to move. Specifically when the mobile

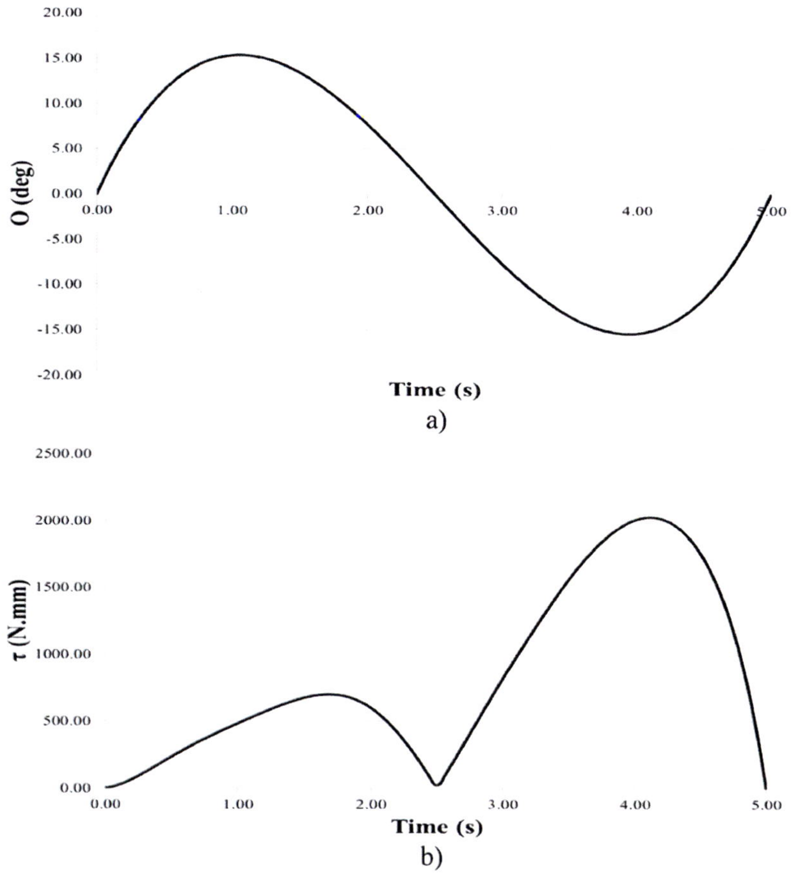

Fig. 11. Data of the driving motor for simulated operation of the joint mechanism solution in Figs. 3: (a) Input angle; (b) Computed torque during simulation in Fig. 7.

platform changes the direction of the motion. Considering all these issues the second version of the joint mechanism in Fig. 16 was designed and built.

Figure 16 represents the second solution chosen as the final design that evolved from the one in Fig. 15 by adopting a kinematic inversion of the joint mechanism to increase the efficiency of transmission of motion and force to the joint platform and to simplify the connection to the antagonist actuating cables. This prototype was built in a version with the main idea of the implementation directly in the TORVEASTRO limbs as shown in Fig. 16 with uses the revolute connector between the arm and forearm. A powerful actuator is expected to actuate this joint mechanism. The maximum torque of the actuator is 7.3 N m when the voltage of the servomotor is 14.8 V and the current is 5.2 A. This servomotor will be used to actuate this design because they are used in TORVEASTRO as the main actuators. In this second version of the joint mechanism the singularities could be present as is shown in Fig. 3. To avoid the singularity, the ratio of the motion transmission is increased using a pulley to connect the servomotor with the mechanism by cables and the size of the links l_1, l_2, l_3, and l_4 as in Table 1.

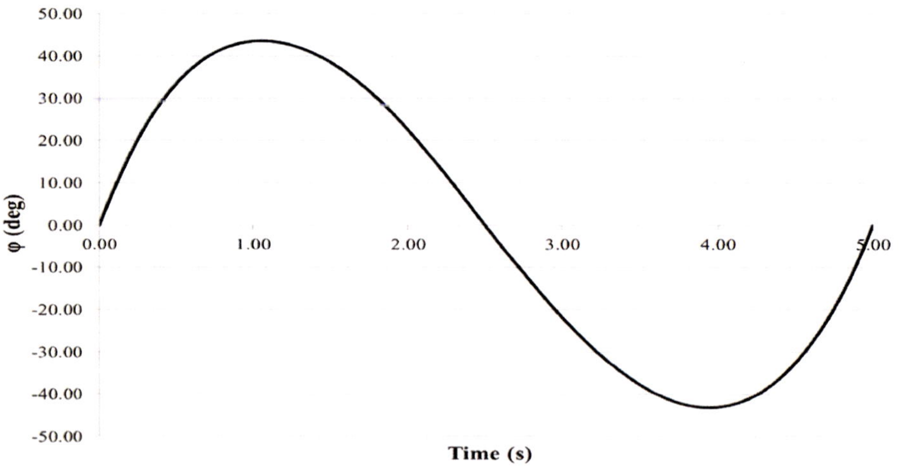

Fig. 12. Computed results of the joint platform simulation in Fig. 7 in terms of angular displacement.

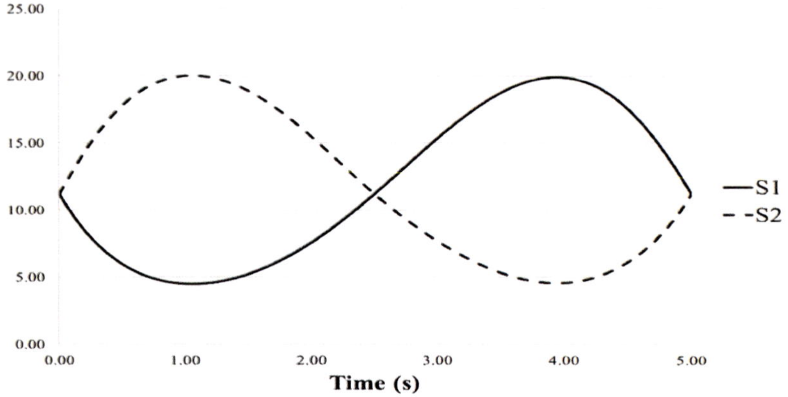

Fig. 13. Computed results of simulation of the strokes S_1 and S_2 in Fig. 7 in terms of linear displacement.

7 Conclusions

The joint mechanisms were designed with cable actuation with a compact efficient operation that could be useful for TORVEASTRO. The second joint mechanism was chosen because the simulated results showed that the efficiency of the transmission increased in comparison to the first joint mechanism design. Furthermore, the gain of the transmission between the actuator and the motion range of the joint mechanism's mobile platform is better than the first prototype because of the kinematic inversion made it. The second joint mechanical design with an overall size of 60 x 35 mm and its operation are characterized through a CAD design that is simulated successfully for a joint range of 86°. Thanks to a torque range. The maximum torque acquired by computed results was 2029.86 N.mm.

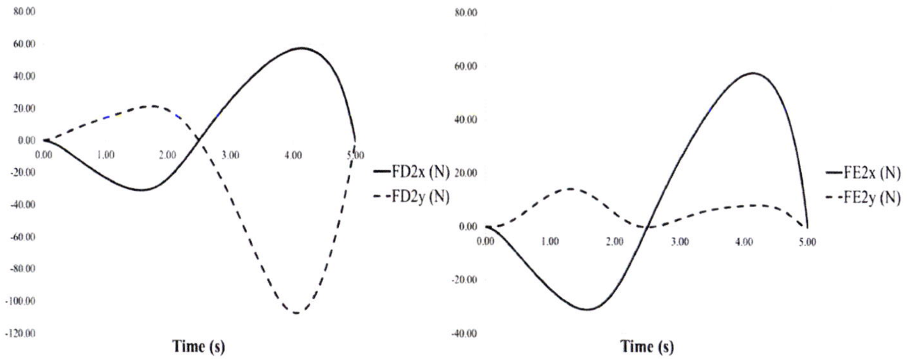

a) b)

Fig. 14. Computed results of the simulation in Fig. 7 in terms of reaction force: (a) $\mathbf{F_{D2x}}$ and $\mathbf{F_{D2y}}$; (b) $\mathbf{F_{E2x}}$ and $\mathbf{F_{E2y}}$.

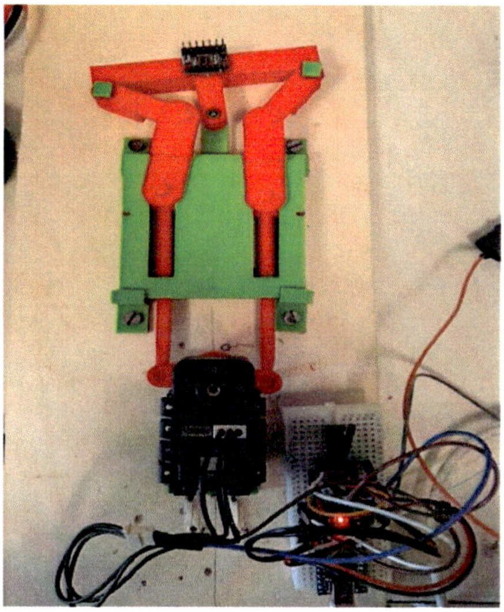

Fig. 15. Prototype of the first proposed joint mechanism in Fig. 2.

Fig. 16. Second proposed joint mechanism layout.

Acknowledgements. This work has been supported by grant Lazioinnova-POR214–2020-A0375–2020-36517 for the project TORVEASTRO which is gratefully acknowledged.

References

1. Jung, Y., Bae, J.: An asymmetric cable-driven mechanism for force control of exoskeleton systems. Mechatronics **40**, 41–50 (2016). https://doi.org/10.1016/j.mechatronics.2016.10.013
2. Hong, H., Ali, J., Ren, L.: A review on topological architecture and design methods of cable-driven mechanism. Adv. Mech. Eng. **10**(5), 1–14 (2018). https://doi.org/10.1177/1687814018774186
3. Liu, T., Mu, Z., Wang, H., Xu, W., Li, Y.: A Cable-driven redundant spatial manipulator with improved stiffness and load capacity. IEEE International Conference Intelligent Robotics System, pp. 6628–6633 (2018). https://doi.org/10.1109/IROS.2018.8593679
4. Li, Z., Chen, W., Zhang, J., Li, Q., Wang, J., Fang, Z., Yang, G.: A novel cable-driven antagonistic joint designed with variable stiffness mechanisms. Mechanism and Machine Theory, p. 171, 104716 (2022). https://doi.org/10.1016/j.mechmachtheory.2021.104716
5. Cho, K.H., Song, M.G., Jung, H., Yang, S.Y., Rodrigue, H., Moon, H., Choi, H.R.: Biomimetic robotic joint mechanism driven by soft linear actuators. IEEE International Conference on Robotics and Automation (ICRA), pp. 1850–1855 (2017). https://doi.org/10.1109/ICRA.2017.7989216
6. Jiang, S., Hua, D., Wang, Y., Ju, F., Yin, L., Chen, B.: Design and modeling of motion-decoupling mechanism for cable-driven joints. Adv. Mech. Eng. **10**(5), 1 (2018). https://doi.org/10.1177/1687814018777428
7. Gosselin, C., Cardou, P., Bruckmann, T., Pott, A. (eds.): Cable-Driven Parallel Robots. MMS, vol. 53. Springer, Cham (2018). https://doi.org/10.1007/978-3-319-61431-1
8. Pott, A.: Cable-Driven Parallel Robots -Theory and Application, Springer Tracts in Advanced Robotics, Springer, Cham (2018). https://doi.org/10.1007/978-3-319-76138-1

9. Pott, A., Bruckmann, T.: Cable-Driven Parallel Robots- Proceedings of the 4th International Conference on Cable-Driven Parallel Robots, Mechanisms and Machine Science Vol. 74, Springer, Cham (2019). https://doi.org/10.1007/978-3-030-20751-9
10. Ceccarelli, M., Russo, M., Morales-Cruz, C.: Parallel architectures for humanoid robots, robotics **9**, 75 (2020). https://doi.org/10.3390/robotics9040075
11. Araque Isidro, J.E., Cafolla, D., Ceccarelli, M.: Problems and Requirements for Outer Space Astronaut Service Robot, Advances in Italian Mechanism Science. IFToMM Italy 2022. Mechanisms and Machine Science, Springer, Cham, vol. 122, pp. 603–611 (2022). https://doi.org/10.1007/978-3-031-10776-4_69
12. Araque Isidro, J.E., Cafolla, D., Ceccarelli, M.: Problems and requirements for outer space astronaut service robot. The 4th IFToMM ITALY Conference (IFIT 2022), September 7–9, Naples, Italy, Paper No. 012 (2022).https://doi.org/10.1007/978-3-031-10776-4_69
13. Ceccarelli, M., Cafolla, D., Araque, J.: Mechanism-driven joint, patent pending no. IT102022000007889, Italy, 1/04/2022
14. Ceccarelli, M.: Fundamentals of Mechanics of Robotic Manipulation, 2nd edn. Springer, Cham (2022). https://doi.org/10.1007/978-3-030-90848-5

Kinematics Equations for the Control System of the TORVEastro Robot

Marco Paoloni(✉), Marco Santoro, and Giacomo Cupertino

ENEA–Italian National Agency for New Technologies, Energy and Sustainable Economic Development–Robotics and Artificial Intelligence (RIA) Lab, 00123 Roma, Italy
{marco.paoloni,marco.santoro,giacomo.cupertino}@enea.it

Abstract. In this paper are described the main kinematics issues of the TORVEastro robot, a new device proposed as service robot for the surveillance and the maintenance of the orbiting space stations. After a short description of its mechanical structure, the forward and the inverse kinematics of its arms are addressed and resolved for manipulation and locomotion tasks. Since the found solutions were implemented as functions for the robot control system software, the method used to test the correct implementation and the obtained results are presented in the conclusions of the paper; the same section also suggests possible applications of the formulated algorithm to the control of simultaneous and coordinated use of two arms.

Keywords: Robotic manipulation · Forward kinematics · Inverse kinematics · Control systems · Space robotics · TORVEASTRO robot

1 Introduction

The TORVEastro research project aims to the development of a new service robot that could assist astronauts in Extra-Vehicular-Activities (EVA) for surveillance and maintenance of the external structures of orbiting space stations, like the International Space Station (ISS). The employment of humans in space requires complex life support systems due to the harsh characteristics of such an environment like high vacuum, large temperature differences, and high ionizing radiations.

The need for complex support and life-saving systems makes astronauts space operations of dangerous and expensive, so their replacement with robots would reduce the risks and costs of surveillance and maintenance activities.

The project starts from a patented concept of a robot constituted by a central body where three identical arms are mechanically connected ([1, 2]); all the actuators are placed in the central body and their movements and forces are transferred to the arms by cable transmissions. The end-effector of each arm can perform manipulation tasks but can also execute the displacements of the whole robot grabbing the rods and the handrails provided outside a space station to help astronauts in their external activities.

The symmetrical assembly of the arms makes them interchangeable; the number of arms was set at three, assuming that this was the minimum necessary to accomplish

© The Author(s), under exclusive license to Springer Nature Switzerland AG 2023
M. Ceccarelli et al. (Eds.): TORVEASTRO 2023, MMS 130, pp. 60–74, 2023.
https://doi.org/10.1007/978-3-031-28447-2_5

dexterous coordinated multi-arms tasks and that a larger number of arms was not appropriate for the increased total weight (it is very difficult to transport objects of large volume or weight from the ground to a space station) and for the greater complexity of the transmission of movements and forces from the body to the arms.

Placing actuators, controllers, drivers, and sensors inside the main body will provide adequate protection to resist such electronic and mechanical devices outside the space station from the space station's ionizing radiation and severe temperature changes.

Each arm of TORVEastro is actuated by three motors connected to three revolute joints, this choice was also adopted to keep the total weight and volume of the proposed robot low.

The actuation of an end-effector by only three motors limits its number of degrees of freedom to three and thus implies three constraints on the variability of the positions and orientations it can achieve. Moreover, to prevent loss of the robot, there must always be at least one end-effector firmly gripped to a support of the orbiting station; actuations of the arm motors of this end-effector will change the poses of the central robot body and the other two end-effectors.

In this way, a free end-effector will be able to perform manipulation tasks by combining the three actuators of its arm with the other three of the arm with the gripping end-effector in a coordinated operation mode with six degrees of freedom.

This way of using two under actuated arms is similar to the one used by Canadarm2 ([3,4]) of the Canadian Space Agency (CSA) and ERA [5] of the European Space Agency (ESA), the two large robotics manipulators currently in use on the ISS to move supplies, equipment, and to catch visiting vehicles and berthing them to the orbiting station. These two devices consist of two long articulated booms (17 m for Canadarm2 when the two arms are fully extended and 11.7 m in the same condition for ERA) connected by an actuated joint.

While these two manipulators have seven degrees of freedom (three for each of the two arms and one for the joint that connects them), a pair of TORVEastro arms working together in a service task has only six; moreover, its revolute joints have a different axis orientation than those servicing ISS.

The study of the kinematic properties of the proposed robot is then an original topic and necessary topic to optimize its use as a future service robot.

In this paper are described the more recent results in the analysis of the robot kinematics (a first approach can be found in [6]) and their implementation as functions of the control system algorithms; after a preliminary description of the robot mechanical structure, exposed in the second section, the TORVEastro arm is modeled in the third one and the forward and the inverse kinematics are discussed in the remaining sections.

2 Robot Description

As depicted in Fig. 1, the TORVEastro robot can be described as a cylindrical central body, on which three identical arms, arranged symmetrically, are mounted; each of the three arms, being made up of three rigid mechanical links, each articulated by a revolute joint, has three degrees of freedom and is equipped with a gripper as end-effector.

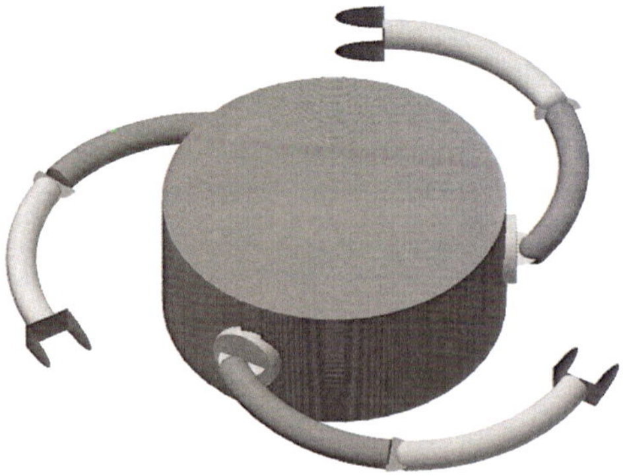

Fig. 1. Sketch of the TORVEastro robot

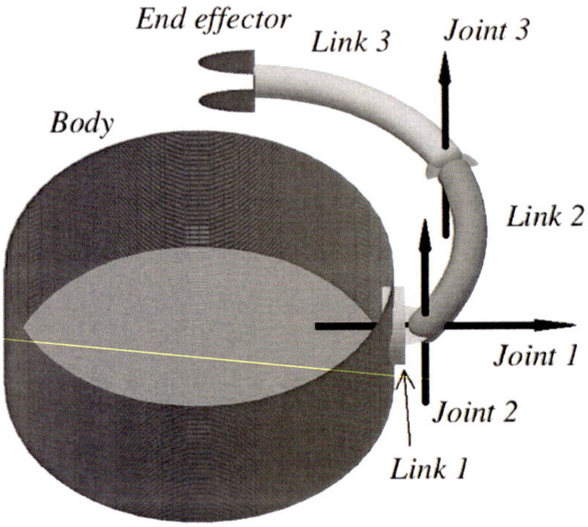

Fig. 2. Kinematic structure of the arms

As depicted in Fig. 2 (where the central body was sketched without its upper surface), each TORVEastro arm consists of an open serial kinematic chain, where the first link is connected to the central body by the first joint, the second link is connected to the first one by the second joint and the third link is connected to the second one by the third joint. The first joint allows the first link to rotate around an axis perpendicular to the body lateral surface, the second joint allows the second link to revolute around an axis intersecting perpendicularly the first joint one and the third joint allows the third link to rotate around an axis parallel to the second joint one (and then perpendicular to the first joint axis); at the end of the chain, the end-effector is rigidly connected to the last link without any other joint.

During the arm movements, the orthogonality of the first and second joint axes and the parallelism of the second and third are always conserved; the intersection point between the first and second axes does not change its position relative to the central body.

3 Preliminary Definitions

For the formulation of the TORVEastro kinematic equations, a coordinate reference frame is embedded into the robot central body as shown in Fig. 3 (where only the side surface of the body is sketched); since all three rotation axes of the first joint of each arm lie on the same plane, the origin is placed on this common plane at the intersection of three axes; after assigning progressive numbering to the arms, the x_B axis is placed on the same plane, directed as the rotation axis of the first joint of the first arm, the z_B axis is directed as the normal to the common plane with its pointing direction chosen in order to see the three arms numbered counterclockwise and the y_B axis is disposed so that this reference frame is right-handed.

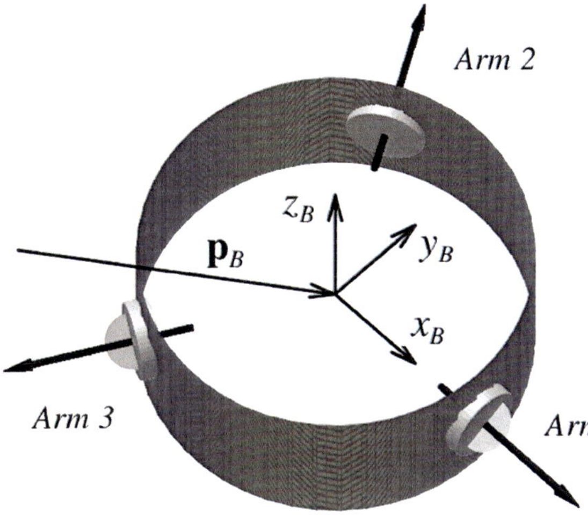

Fig. 3. Body reference frame

To describe the position of the central body in its working environment (in the aim of the project, an orbiting space station), the point vector $\mathbf{p}_B$ is adopted, which expresses the position of the origin of the body frame in a coordinate reference frame fixed on the station; in this last frame, the body orientation will be expressed by the three unit vectors $\mathbf{x}_B$, $\mathbf{y}_B$ and $\mathbf{z}_B$ relative to the body frame axes. These four vectors are used to define a 4 x 4 homogeneous matrix $\mathbf{Q}_B$ (the body pose matrix) expressed by.

$$\mathbf{Q}_B = \begin{bmatrix} \mathbf{x}_B & \mathbf{y}_B & \mathbf{z}_B & \mathbf{p}_B \\ 0 & 0 & 0 & 1 \end{bmatrix} \tag{1}$$

The matrix $\mathbf{Q}_B$ can also be used to transform the coordinates of a point referred to the body reference frame in the corresponding coordinates referred to the space station.

Other coordinate reference frames are placed on the three end-effectors to describe their positions and orientations. In Fig. 4, their axes are directed as follows: one axis lies along the direction from which the gripper would approach an object, another axis is in the direction specifying the orientation of the gripper (from a finger to the other) and the

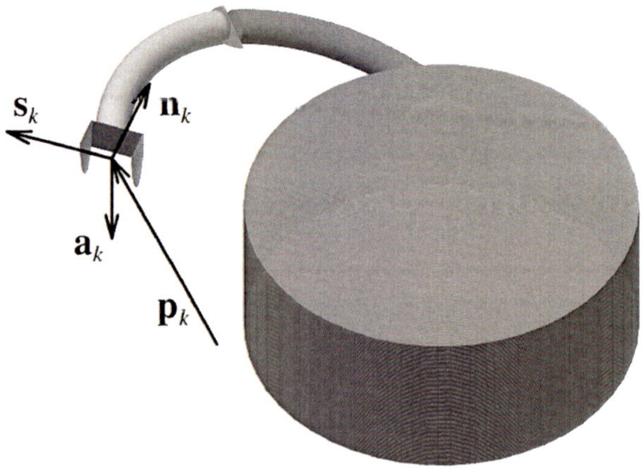

Fig. 4. End-effector reference frame

last axis must form a right-handed reference frame: associated with these axes, for the k-th arm, are the unit vectors $\mathbf{a}_k$, $\mathbf{s}_k$ and $\mathbf{n}_k$, respectively, whose components are referred to the station reference frame. The origins of these reference frames will describe the position of the end-effectors; for the k-th arm, this origin will be described, in the station frame, by the vector $\mathbf{p}_k$.

For some tasks, the required pose of the end-effector can be referred to the body reference frame, such as when an operation is supervised by a camera on board the TORVEastro itself; it is therefore useful to refer the set of four vectors associated with the end-effector pose to this frame as well. For this purpose, the vectors ${}^B\mathbf{n}_k$, ${}^B\mathbf{s}_k$, ${}^B\mathbf{a}_k$ and ${}^B\mathbf{p}_k$ are introduced, where the superscript indicates that they are referred to the reference frame of the central body.

These two sets of four vectors are used to define two different versions of the end-effector pose matrix of the k-th arm, one referring to the station reference frame and the other referring to the body reference frame, expressed by

$$\mathbf{Q}_k = \begin{bmatrix} \mathbf{n}_k & \mathbf{s}_k & \mathbf{a}_k & \mathbf{p}_k \\ 0 & 0 & 0 & 1 \end{bmatrix} \; {}^B\mathbf{Q}_k = \begin{bmatrix} {}^B\mathbf{n}_k & {}^B\mathbf{s}_k & {}^B\mathbf{a}_k & {}^B\mathbf{p}_k \\ 0 & 0 & 0 & 1 \end{bmatrix} \quad (2)$$

Each of the two pose matrices can be transformed into the other using the body pose matrix $\mathbf{Q}_B$ as in the following expressions.

$$\mathbf{Q}_k = \mathbf{Q}_B \cdot {}^B\mathbf{Q}_k \; \mathbf{Q}_k = {}^B\mathbf{Q}_B^{-1} \cdot \mathbf{Q}_k \quad (3)$$

A task assigned to the k-th arm can be formulated as a sequence of orientations and positions to be assumed by its end-effector by appropriately articulating its links; therefore, it is very important to formulate the kinematic equations expressing the dependence of the end-effector's pose on the rotations around the axes. To describe the rotation around the j-th axis of the k-th arm, the joint angle ϑ_{jk} is adopted, considered positive for counterclockwise rotations referring to the pointing directions of the rotation axes defined in Fig. 2.

4 Arm Kinematics

4.1 Forward Kinematics

To solve the arm forward kinematics, the end-effector pose matrix must be expressed as function of the joint angles. To accomplish this task for every open serial kinematic chain, a general algorithm (described in [7] and [8]) has been developed, which makes use of Denavit and Hartenberg notation [9] to prescribe the appropriate construction of a set of coordinate reference frames (one for each link in the chain) and for each frame of the set evaluates the homogeneous matrix that performs the transformation of the components of a homogeneous vector referred to that frame into the corresponding ones referred to the frame embedded in the previous link.

To apply this algorithm to the TORVEastro robot arms, three auxiliary body reference frames are adopted, one for each arm (see Fig. 5); for the k-th arm, the origin of the auxiliary frame is placed at the intersection of the rotation axes of the first and second joints, the z_{0k} axis is aligned with the rotation axis of the first joint pointing outward from the body, the y_{0k} axis is aligned with the z_B axis and the x_{0k} axis is chosen in order to obtain a right-handed reference frame.

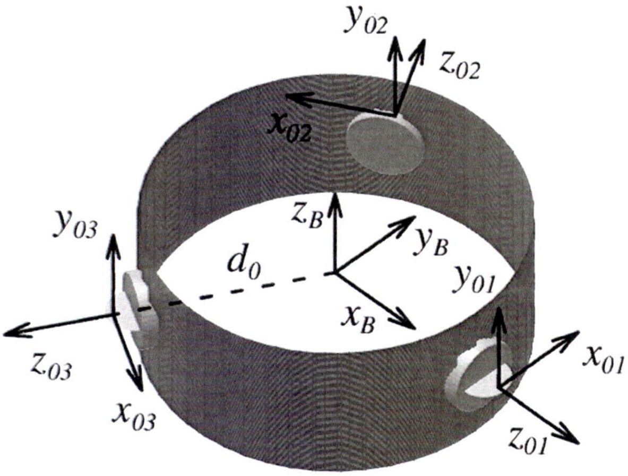

Fig. 5. Auxiliary body reference frames

During arms movements, the positions and orientations of the auxiliary frames remain fixed in the body reference frame, the common distance between their origins and the origin of the main body frame is denoted by d_0.

After the auxiliary one, another reference frame is fixed on the first link of the k-th arm. As it can be seen in Fig. 6, its origin is the same as for the auxiliary frame, the z_{1k} axis is the rotation axis of the second joint and the x_{1k} axis is aligned with the common normal to the z_{1k} and z_{0k} axes; the positive direction of the x_{1k} axis is chosen so that its unit vector can be obtained from the cross product of those of the z_{1k} and z_{0k} axes. Finally, the y_{1k} axis is disposed to obtain a right-handed reference frame.

As it can be seen in Fig. 7, for the reference frame embedded in the second link of the k-th arm, the origin is placed on the intersection of the third joint axis with its

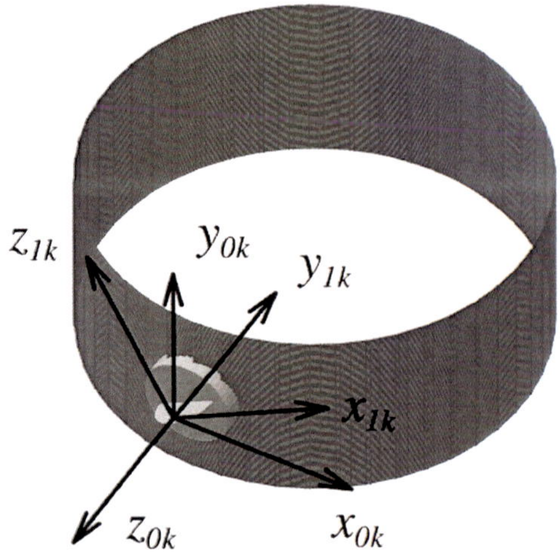

Fig. 6. First link reference frame

normal passing through the origin of the frame embedded in the first link; the z_{2k} axis is the rotation axis of the third joint and the x_{2k} axis is placed along the quoted normal with positive direction from z_{1k} to z_{2k} and finally the y_{2k} axis is arranged to obtain a right-handed reference frame. The distance between the origins of the frames embedded in the first and second links is denoted by a_2.

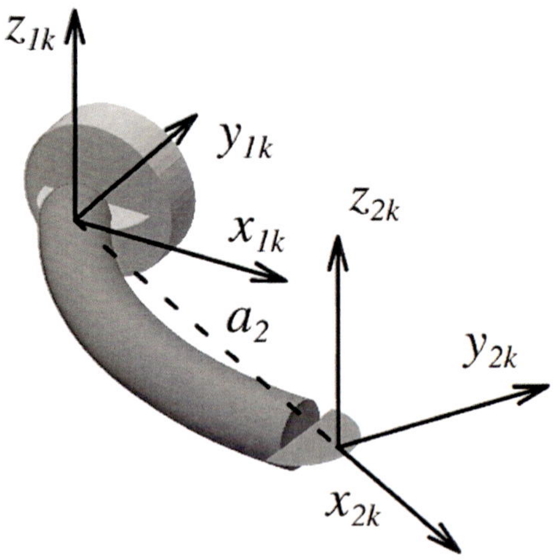

Fig. 7. Second link reference frame

Finally, the reference frame embedded in the third link of the k-th arm is defined (see Fig. 8); its z_{3k} axis is aligned with the end-effector approach direction and its x_{3k} axis is aligned with the normal to the z_{3k} axis passing through the origin of the frame embedded in the second link; the origin of this frame is then placed at the intersection of the x_{3k} and the z_{3k} axes and the positive direction of the x_{3k} axis is from the origin of the

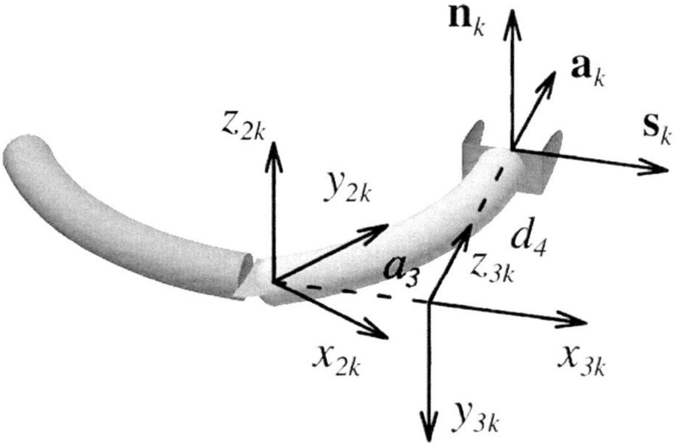

Fig. 8. Third link reference frame

second link frame to that of the third; the y_{3k} axis is arranged to obtain a right-handed reference frame.

The distance between the origin of the frame embedded in the second link and the origin of the frame embedded in the third link is denoted by a_3, while the distance between the same origin and the origin of the frame embedded in the end-effector is denoted by d_4.

After the assignments of these reference frames, it will be possible to formulate the homogeneous matrices expressing the transformation of the components of a homogeneous vector referred to a frame into the ones referred to the previous frame. It must be noted that, according to the Denavit and Hartenberg notation, the reference frames of the links have been expressly assigned so that the $z_{(j-1)k}$ axis is the axis of the j-th joint of the k-th arm and the joint angle ϑ_{jk} is defined as the angle from the $x_{(j-1)k}$ axis to the x_{jk} one and it is considered positive for the counterclockwise verse around this axis.

The transformation from the k-th arm auxiliary body reference frame to the main body reference frame will then be performed by the $\mathbf{S}_k$ matrix expressed by.

$$\mathbf{S}_k = \begin{bmatrix} -\sin\beta_k & 0 & \cos\beta_k & d_0\cos\beta_k \\ \cos\beta_k & 0 & \sin\beta_k & d_0\sin\beta_k \\ 0 & 1 & 0 & 0 \\ 0 & 0 & 0 & 1 \end{bmatrix} \tag{4}$$

where the parameter β_k will have the value 0° for the first arm, the value 120° for the second and the value 240° for the third.

The transformation from the reference frame embedded in the j-th link of the k-th arm to the previous of the same arm can be performed by the matrix $\mathbf{A}_j(\vartheta_{jk})$ expressed by

$$\mathbf{A}_1(\vartheta_{1k}) = \begin{bmatrix} \cos\vartheta_{1k} & 0 & -\sin\vartheta_{1k} & 0 \\ \sin\vartheta_{1k} & 0 & \cos\vartheta_{1k} & 0 \\ 0 & -1 & 0 & 0 \\ 0 & 0 & 0 & 1 \end{bmatrix}$$

$$\mathbf{A}_2(\vartheta_{2k}) = \begin{bmatrix} \cos\vartheta_{2k} & -\sin\vartheta_{2k} & 0 & a_2\cos\vartheta_{2k} \\ \sin\vartheta_{2k} & \cos\vartheta_{2k} & 0 & a_2\sin\vartheta_{2k} \\ 0 & 0 & 1 & 0 \\ 0 & 0 & 0 & 1 \end{bmatrix}$$

$$\mathbf{A}_3(\vartheta_{3k}) = \begin{bmatrix} \cos\vartheta_{3k} & 0 & -\sin\vartheta_{3k} & a_3\cos\vartheta_{3k} \\ \sin\vartheta_{3k} & 0 & \cos\vartheta_{3k} & a_3\sin\vartheta_{3k} \\ 0 & -1 & 0 & 0 \\ 0 & 0 & 0 & 1 \end{bmatrix} \tag{5}$$

Finally, the transformation from the end-effector frame to the third link frame can be performed by the **E** matrix (the same for all arms) expressed by.

$$\mathbf{E} = \begin{bmatrix} 0 & 1 & 0 & 0 \\ -1 & 0 & 0 & 0 \\ 0 & 0 & 1 & d_4 \\ 0 & 0 & 0 & 1 \end{bmatrix} \tag{6}$$

It should be noted that, in the notation assigned in the introduction of these matrices, the $\mathbf{S}_k$ matrix is a constant matrix that does not depend on the joint angles, but is different for each arm; with the notation $\mathbf{A}_j\left(\vartheta_{jk}\right)$ is denoted a matrix function $\mathbf{A}_j(\vartheta)$ typical of the reference frame associated to the j-th link, identical in its formulation for all arms (thus devoid of arm index) and taking different values only when evaluated for the joint angle ϑ_{jk} of a given arm; finally, the **E** matrix is a constant matrix identical for all arms, thus devoid of any kind of index.

With the use of these five matrices and of the matrix $\mathbf{Q}_B$, it is possible to transform the coordinates of a homogeneous vector referred to the frame embedded in the end-effector of the k-th arm reference frame into the corresponding ones, referred to the station reference frame by means of a chain of transformations; these components are first transformed into the corresponding ones referred to the frame embedded in the third link of the same arm frame using the matrix **E**, then the obtained components are transformed in to the ones referred to the frame embedded in the second link using the matrix $\mathbf{A}_3(\vartheta_{3k})$. The components referred to the frame embedded in the second link are then obtained by the transformation performed by the matrix $\mathbf{A}_2(\vartheta_{2k})$ and similarly the components referred to the frame embedded in the first link are then transformed into the ones referred to the auxiliary body reference frame by the $\mathbf{A}_1(\vartheta_{1k})$ matrix. The components referred to the auxiliary body frame are transformed into those referred to the main reference frame of the body by the $\mathbf{S}_k$ matrix and, finally, the latter are transformed into the corresponding ones expressed in the station frame by the $\mathbf{Q}_B$ matrix.

The coordinates transformation from the frame embedded in the end-effector to the frame embedded in the central body is also performed directly by the pose matrix $\mathbf{Q}_k$, which then must be equal to the ordered product of the five introduced matrices; this yields the expression

$$\mathbf{Q}_k(\vartheta_{1k}, \vartheta_{2k}, \vartheta_{3k}) = \mathbf{Q}_B \cdot \mathbf{S}_k \cdot \mathbf{A}_1(\vartheta_{1k}) \cdot \mathbf{A}_2(\vartheta_{2k}) \cdot \mathbf{A}_3(\vartheta_{3k}) \cdot \mathbf{E} \tag{7}$$

Noting that, according to Eqs. (1) and (4), the product $\mathbf{Q}_B \cdot \mathbf{S}_k$ can be written as

$$\mathbf{Q}_B \cdot \mathbf{S}_k = \begin{bmatrix} \mathbf{x}_{0k} & \mathbf{y}_{0k} & \mathbf{z}_{0k} & \mathbf{p}_B + d_0\mathbf{z}_{0k} \\ 0 & 0 & 0 & 1 \end{bmatrix} \tag{8}$$

where $\mathbf{x}_{0k}$, $\mathbf{y}_{0k}$ and $\mathbf{z}_{0k}$ are the unit vectors associated with the k-th arm auxiliary reference frame, expressed by

$$\mathbf{x}_{0k} = -\sin\beta_k\mathbf{x}_B + \cos\beta_k\mathbf{y}_B \; \mathbf{y}_{0k} = \mathbf{z}_B \; \mathbf{z}_{0k} = \cos\beta_k\mathbf{x}_B + \sin\beta_k\mathbf{y}_B \tag{9}$$

By performing the other products in Eq. (7), it is then possible to obtain a closed form solution of the arm forward kinematics by means of the expressions

$$\begin{aligned} \mathbf{n}_k &= -S\vartheta_{1k}\mathbf{x}_{0k} + C\vartheta_{1k}\mathbf{y}_{0k} \\ \mathbf{s}_k &= C\vartheta_k(C\vartheta_{1k}\mathbf{x}_{0k} + S\vartheta_{1k}\mathbf{y}_{0k}) - S\vartheta_k\mathbf{z}_{0k} \\ \mathbf{a}_k &= -S\vartheta_k(C\vartheta_{1k}\mathbf{x}_{0k} + S\vartheta_{1k}\mathbf{y}_{0k}) - C\vartheta_k\mathbf{z}_{0k} \\ \mathbf{p}_k &= \mathbf{p}_B + (a_2C\vartheta_{2k} + a_3C\vartheta_k - d_4S\vartheta_k)(C\vartheta_{1k}\mathbf{x}_{0k} + S\vartheta_{1k}\mathbf{y}_{0k}) \\ &\quad + (d_0 - a_2S\vartheta_{2k} - a_3S\vartheta_k - d_4C\vartheta_k)\mathbf{z}_{0k} \end{aligned} \tag{10}$$

where the terms $S\vartheta_{jk}$ and $C\vartheta_{jk}$ were introduced to denote the sine and the cosine of the joint angle ϑ_{jk} and the terms $S\vartheta_k$ and $C\vartheta_k$ to denote the sine and the cosine of the sum $\vartheta_{2k} + \vartheta_{3k}$.

Using the first three expressions of the unit vectors in (10), the last expression can also be rewritten as

$$\mathbf{p}_k = \mathbf{p}_B + d_0\mathbf{z}_{0k} + (a_2C\vartheta_{3k} + a_3)\mathbf{s}_k + (-a_2S\vartheta_{3k} + d_4)\mathbf{a}_k \tag{11}$$

4.2 Inverse Kinematics

In order to solve the arm inverse kinematics, it is necessary to evaluate the joint angles of the k-th arm that allow its end-effector to assume a desired pose expressed by the assignment of the four vectors $\mathbf{n}_k$, $\mathbf{s}_k$, $\mathbf{a}_k$, and $\mathbf{p}_k$, when the central body has assumed the pose expressed by the other four vectors $\mathbf{x}_B$, $\mathbf{y}_B$, $\mathbf{z}_B$, and $\mathbf{p}_B$. Therefore, the first three expressions in Eq. (10) and the expression in Eq. (11) should be considered as equations in the unknown quantities represented by the joint angles.

From a first look at the first expression of Eq. (10), it is possible to state that the unit vector $\mathbf{n}_k$ will be orthogonal to the unit vector $\mathbf{z}_{0k}$ for any given configuration of the joint angles; this is easily explained considering that the vector $\mathbf{n}_k$ is parallel to the third joint axis (and thus also to the second), while the vector $\mathbf{z}_{0k}$ is aligned to the first joint axis, which is perpendicular to the second.

A second constraint can be recognized in Eq. (11), where, because of the first constraint, it can be stated that the unit vector $\mathbf{n}_k$ will always be orthogonal to the vector $\mathbf{p}_k - \mathbf{p}_B$ as well.

The third and final constrain can be obtained by rewriting Eq. (11) as

$$\mathbf{p}_k - \mathbf{p}_B - d_0\mathbf{z}_{0k} - a_3\mathbf{s}_k - d_4\mathbf{a}_k = a_2(C\vartheta_{3k}\mathbf{s}_k - S\vartheta_{3k}\mathbf{a}_k) \tag{12}$$

and noting that this expression implies the condition.

$$|\mathbf{p}_k - \mathbf{p}_B - d_0\mathbf{z}_{0k} - a_3\mathbf{s}_k - d_4\mathbf{a}_k| = a_2 \tag{13}$$

After a valid assignment of the end-effector pose, to accomplish the arm inverse kinematics of the arm, it is possible to obtain from the first three expressions of Eq. (10) the four expressions.

$$\begin{cases} \cos\vartheta_{1k} = \mathbf{n}_k \cdot \mathbf{y}_{0k} \\ \sin\vartheta_{1k} = -\mathbf{n}_k \cdot \mathbf{x}_{0k} \end{cases} \begin{cases} \cos(\vartheta_{2k} + \vartheta_{3k}) = -\mathbf{a}_k \cdot \mathbf{z}_{0k} \\ \sin(\vartheta_{2k} + \vartheta_{3k}) = -\mathbf{s}_k \cdot \mathbf{z}_{0k} \end{cases} \tag{14}$$

which allow to obtain univocally the angle ϑ_{1k} and the sum of the angles ϑ_{2k} and ϑ_{3k}.

Considering Eq. (11), it is possible to write the two expressions.

$$\begin{cases} a_2\cos\vartheta_{3k} = (\mathbf{p}_k - \mathbf{p}_B - d_0\mathbf{z}_{0k}) \cdot \mathbf{s}_k - a_3 \\ a_2\sin\vartheta_{3k} = d_4 - (\mathbf{p}_k - \mathbf{p}_B - d_0\mathbf{z}_{0k}) \cdot \mathbf{a}_k \end{cases} \tag{15}$$

which allow to obtain uniquely the angle ϑ_{3k} and then the angle ϑ_{2k} from the value found for the sum $\vartheta_{2k} + \vartheta_{3k}$ using Eq. (14).

The solution obtained from the assignment of a well constrained end-effector pose is unique, but its feasibility must be verified in order to be accepted; it is possible that the angle joints found may lead a link too close to the central body (or even inside, as in Fig. 9).

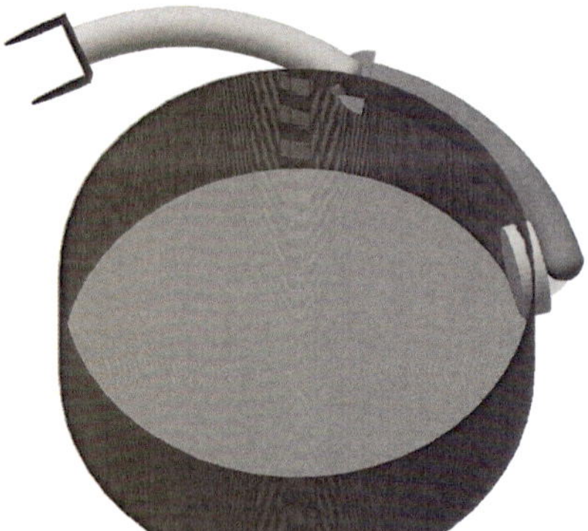

Fig. 9. Arm-body interference

For this reason, safety acceptance limits have been defined: safe values for the angle ϑ_{1k} can range from -180° to 180°, for the angle ϑ_{2k} from $-150°$ to 30° and for the sum $\vartheta_{2k} + \vartheta_{3k}$ from $-120°$ to 60°. In a forthcoming development, the robot control system will include an interference and collision avoidance algorithm that, for a given set of joint angles, will evaluate the distances between the links and the central body to accept only those solutions that will have these distances greater than a safe value.

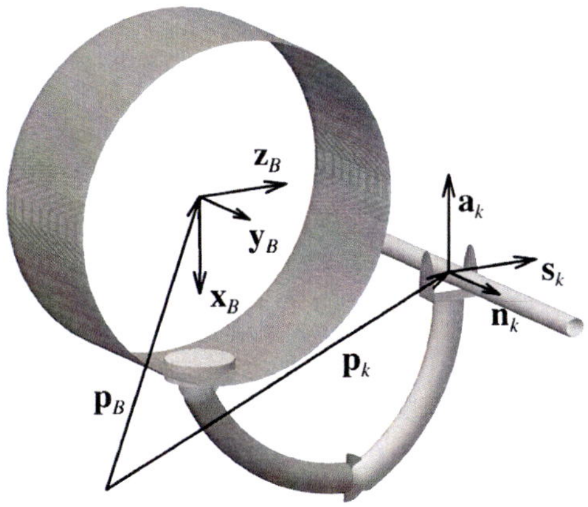

Fig. 10. Body forward kinematics

5 Body Kinematics

The arms of the TORVEastro robot do not only act as manipulators; when a gripper firmly grasps a handrail used by astronauts in extravehicular activities, a change in the joint angles of its arm will result in a change in the position and orientation of the robot central body (see Fig. 10). Also in this case, the kinematic chain from the fixed end-effector to the moving central body is an open serial kinematic chain that could be described by the Denavit and Hartenberg notation, but the reference systems to be adopted would be different from those already introduced. To simplify the development of the control system software, the introduced frames have been maintained.

5.1 Forward Kinematics

To solve the body forward kinematics problem, it is necessary to obtain the body pose matrix $\mathbf{Q}_B$ from the assignment of the end-effector pose matrix $\mathbf{Q}_k$ and the joint angles of the k-th arm. The two poses will be expressed in the station reference frame.

By rearranging Eq. (7), the solution for the body forward kinematics can be formally expressed by

$$\mathbf{Q}_B(\vartheta_{1k}, \vartheta_{2k}, \vartheta_{3k}) = \mathbf{Q}_k \cdot \mathbf{E}^{-1} \cdot \mathbf{A}_3^{-1}(\vartheta_{3k}) \cdot \mathbf{A}_2^{-1}(\vartheta_{2k}) \cdot \mathbf{A}_1^{-1}(\vartheta_{1k}) \cdot \mathbf{S}_k^{-1} \quad (16)$$

Without executing the multiplication chain, the unit vectors expressing the orientation of the auxiliary frame of the k-th arm can be obtained directly by the inversion of the first three equations in Eq. (10), obtaining the expressions.

$$\begin{aligned} \mathbf{x}_{0k} &= C\vartheta_{1k}(C\vartheta_k \mathbf{s}_k - S\vartheta_k \mathbf{a}_k) - S\vartheta_{1k}\mathbf{n}_k \\ \mathbf{y}_{0k} &= S\vartheta_{1k}(C\vartheta_k \mathbf{s}_k - S\vartheta_k \mathbf{a}_k) + C\vartheta_{1k}\mathbf{n}_k \\ \mathbf{z}_{0k} &= -\ (S\vartheta_k \mathbf{s}_k + C\vartheta_k \mathbf{a}_k) \end{aligned} \quad (17)$$

From these vectors, the unit vectors of the central body frame can be obtained by inverting the expressions in Eq. (9) to obtain the following final expressions

$$\mathbf{x}_B = -\sin\beta_k \mathbf{x}_{0k} + \cos\beta_k \mathbf{z}_{0k} \ \mathbf{y}_B = \cos\beta_k \mathbf{x}_{0k} + \sin\beta_k \mathbf{z}_{0k} \ \mathbf{z}_B = \mathbf{y}_{0k} \quad (18)$$

To complete the forward kinematics, to obtain the vector $\mathbf{p}_B$, Eq. (11) is rewritten as

$$\mathbf{p}_B = \mathbf{p}_k - d_0\mathbf{z}_{0k} - (a_2C\vartheta_{3k} + a_3)\mathbf{s}_k + (a_2S\vartheta_{3k} - d_4)\mathbf{a}_k \tag{19}$$

and, using the last expression in Eq. (17), as

$$\mathbf{p}_B = \mathbf{p}_k + (d_0S\vartheta_k - a_2C\vartheta_{3k} - a_3)\mathbf{s}_k + (d_0C\vartheta_k + a_2S\vartheta_{3k} - d_4)\mathbf{a}_k \tag{20}$$

In Eq. (20), it can be seen that the position of the center of the robot body is independent from the ϑ_{1k} angle; this fact can be easily explained considering that this center is located on the axis of the first joint of each arm and therefore its position does not change for a change in the angle ϑ_{1k}, which only affects the body orientation.

5.2 Inverse Kinematics

In the formulation of the body inverse kinematics problem, a pose matrix $\mathbf{Q}_k$ of the k-th arm end-effector (e.g., the one needed to grasp a handle on the space station) and a pose matrix $\mathbf{Q}_B$ of the central body are assigned and the joint angles ϑ_{1k}, ϑ_{2k} and ϑ_{3k} that allow for the assigned poses must be evaluated; typically the two poses will be expressed in the station reference frame. Again, having only three actuators, the position and orientation of the body will be subject to some constraints; for example, seeing the last expression in Eq. (17), the unit vectors $\mathbf{z}_{0k}$ and $\mathbf{n}_k$ must still be orthogonal.

Seeing Eq. (19), the same happens to the vector to the vector $\mathbf{p}_k - \mathbf{p}_B$, and the same constraint expressed in Eq. (13) can be obtained from Eq. (19).

This is a consequence of the fact that the expressions in Eqs. (10) and (11) are the same as the those in Eqs. (17) and (19), they differ only in the pose made explicit. The two inverse kinematics problems differ only for the first pose arbitrarily assigned (the end-effector pose for the arm inverse kinematics and the body pose for the body inverse kinematics); once assigned, the other pose must be assigned under the common constraints.

In both the two inverse kinematics problems, the unknown quantities are the joint angles; from the previous exposition, it can be understood that the expressions found for the arm inverse kinematics solution, given in Eqs. (14) and (15), also leads to the solutions for the body inverse kinematics.

6 Conclusions

Forward and inverse kinematics of TORVEastro robot arms were fully investigated for both manipulation and locomotion tasks; from the expressions obtained, three software functions were implemented for the control system; the first evaluates the pose of the end-effector of a given arm when its set of joint angles and the pose of the central body are assigned, the second evaluates the pose of the central body when the set of joint angles and the pose of the end-effector of a given arm are assigned, and the third evaluates the set of joint angles of a given arm when the poses of its end-effector and central body are assigned, and also evaluates the numerical values of the constraint functions on the two poses to examine their correct assignment.

To verify the correct behavior of the implemented functions, one million test cycles were run for each of the three arms; in each test cycle, a central body pose and a random set of joint angles were generated; then the first function was used with these assignments to evaluate the pose of the arm under consideration; then this pose and that of the central body pose were used to reconstruct the joint angles randomly assigned by the third function, and finally the end-effector pose and the random set of joint angles were used in the second function to reevaluate the central body pose.

In each cycle, each reconstructed joint angle was compared with its assigned random value to evaluate the reconstruction error, and the central body pose obtained at the end of the cycle was compared with the random one generated at the beginning to evaluate an orientation error and a position error. For the execution of the test, the dimensional parameters d_0 and a_2 were assigned the common value of 195 mm, the parameter a_3 was assigned the value of 97.5 mm, and the parameter d_4 was assigned the value of 170 mm, which are the values of the demonstration prototype made in the development of the research project [10].

At the end of the test, it was found that the maximum reconstruction error on the joint angles was $2{\cdot}10^{-13}$ degrees, the maximum orientation error on the reconstructed central body pose was $3{\cdot}10^{-6}$ degrees and the maximum position error on the same pose was $2{\cdot}10^{-13}$ mm.

Because the poses used to test the function that evaluates the joint angles set, when the end-effector and central body poses are assigned always satisfy the kinematic constraints for their own construction, the correct detection of a constraint violation was tested separately on a set of appropriately constructed pairs of poses.

The two forward kinematics obtained can be composed together to describe the simultaneous and coordinated use of two arms. For example, it can be assumed that the end-effector of one arm has grasped a handrail and the end-effector of another must perform a manipulation task, actuated by the simultaneous variation of the two corresponding sets of joint angles.

Again, from the fixed end-effector to the moving one, an open serial kinematic chain is obtained, but it has six degrees of freedom; thus, the moving end-effector will be able to perform more complex tasks.

Assuming that the fixed end-effector is that of the n-th arm, its pose will be expressed in the station reference frame by the assignment of its pose matrix $\mathbf{Q}_n$, and the pose matrix $\mathbf{Q}_k$ of the k-th arm (the one that is supposed to be moved) will have to be evaluated. By introducing the matrix $\mathbf{T}_3(\vartheta_{1k}, \vartheta_{2k}, \vartheta_{3k})$ defined as

$$\mathbf{T}_3(\vartheta_{1k}, \vartheta_{2k}, \vartheta_{3k}) = \mathbf{A}_1(\vartheta_{1k}) \cdot \mathbf{A}_2(\vartheta_{2k}) \cdot \mathbf{A}_3(\vartheta_{3k}) \tag{21}$$

according to Eq. (16), the body pose obtained from the set of joint angles of the n-th arm will be expressed by

$$\mathbf{Q}_B(\vartheta_{1n}, \vartheta_{2n}, \vartheta_{3n}) = \mathbf{Q}_n \cdot \mathbf{E}^{-1} \cdot \mathbf{T}_3^{-1}(\vartheta_{1n}, \vartheta_{2n}, \vartheta_{3n}) \cdot \mathbf{S}_n^{-1} \tag{22}$$

and, according to Eq. (7), the pose of the moving end-effector will be expressed by

$$\mathbf{Q}_k(\vartheta_{1n}, \vartheta_{2n}, \vartheta_{3n}, \vartheta_{1k}, \vartheta_{2k}, \vartheta_{3k}) = \mathbf{Q}_B(\vartheta_{1n}, \vartheta_{2n}, \vartheta_{3n}) \cdot \mathbf{S}_k \cdot \mathbf{T}_3(\vartheta_{1k}, \vartheta_{2k}, \vartheta_{3k}) \cdot \mathbf{E} \tag{23}$$

By substituting the expression of the central body pose of Eq. (22) in Eq. (23) and developing the product of the matrices, it is possible to obtain a closed form expression of the pose matrix of the moving end-effector as a function of the two set of joint angles; however, during this type of motion, the robot control system must be able to know the pose of the central body to avoid interferences with the space station elements. For this reason, the software function implemented for the solution of this kinematics problem simply uses the sequence of the two forward kinematics functions.

Acknowledgments. The TORVEastro project activities reported in this paper were funded by Regione Lazio [Project 36517 POR FESR LAZIO 2014–20].

References

1. Samani, F., Ceccarelli, M.: Prototype Design and Testing of TORVEastro, Cable-Driven Astronaut Robot. In: Zeghloul, S., Laribi, M.A., Sandoval Arevalo, J.S. (eds.) RAAD 2020. MMS, vol. 84, pp. 448–455. Springer, Cham (2020). https://doi.org/10.1007/978-3-030-48989-2_48
2. Samani, F., Ceccarelli, M.: An experimental characterization of TORVEastro, cable-driven astronaut robot. Robotics **10**(1), 21 (2021). https://doi.org/10.3390/robotics10010021
3. About Canadarm2, CSA webpage, https://www.asc-csa.gc.ca/eng/iss/canadarm2/about.asp. Accessed 17 Jan 2023
4. Remote Manipulator System (Canadarm2), NASA webpage, https://www.nasa.gov/mission_pages/station/structure/elements/remote-manipulator-system-canadarm2/. Accessed 17 Jan 2023
5. European Robotic Arm, ESA webpage, https://www.esa.int/Science_Exploration/Human_and_Robotic_Exploration/International_Space_Station/European_Robotic_Arm. Accessed 17 Jan 2023
6. Paoloni, M., Cupertino, G., Ceccarelli, M.: A kinematic analysis of the TORVEastro astronaut robot arm. In: Proceedings of the SYROM2022 International Conference, Springer Nature Switzerland AG (2020), being published
7. Paul, R.P.: Robot Manipulators: Mathematics Programming and Control. MIT Press, Cambridge (1981)
8. Paul, R.P., Shimano, B., Mayer, G.E.: Kinematic control equations for simple manipulators. IEEE Trans. Syst., Man, Cybern. **11**, 449–455 (1981). https://doi.org/10.1109/TSMC.1981.4308713
9. Denavit, J., Hartenberg, R.S.: A kinematic notation for lower-pair mechanisms based on matrices. J. Appl. Mech. 215–221 (1955)
10. Cafolla, D.: Design and preliminary testing of Torveastro, an outer space service robot. In: Proceedings of the SYROM2022 International Conference, Springer Nature Switzerland AG (2020), being published

Arm Control Design in TORVEastro Prototype

Giacomo Cupertino(✉), Angelo Cipollini, Marco Santoro, and Marco Paoloni

ENEA–Italian National Agency for New Technologies, Energy and Sustainable Economic Development—Robotics and Artificial Intelligence (RIA) Lab, Rome, Italy
{giacomo.cupertino,angelo.cipollini,marco.santoro, marco.paoloni}@enea.it

Abstract. This paper describes the robot arm control design of the TORVEastro project. TORVEastro is a research program dedicated to the development of a new service robot that can assist astronauts in extravehicular activities (EVA) aboard orbiting space stations. After a brief description of the electronic system architecture and its implementation on the laboratory demonstrator robot, the control design is described. Preliminary control tests of a single arm with its basic movements are then illustrated; finally, the main critical aspects and results are reviewed.

Keywords: Control design · Kinematics control · Service robots · Space robotics · Astronaut robots · TORVEastro robot

1 Introduction

The goal of the TORVEastro [2] is to develop a new service robot that can assist astronauts in Extra-Vehicular-Activities (EVA) aboard orbiting space stations; its typical tasks could be surveillance and maintenance of external facilities. The main challenge is to design new mechanisms and solutions for the robot's actuators that conform to the unique requirements of the space environment. Thus, reliability and accuracy are key aspects from the perspective of the robot control system.

After the realization of an initial rough prototype dated 2020 [1] lacking any kind of control, this paper describes the results of the implementation of low-level control algorithms on an higher TRL (Technology Readiness Level) motion robot, limited to a single arm but extendable to the whole robot, according to the equations solved in the paper [3], which discusses the uniqueness of the solution posed in an inverse kinematics problem.

After a preliminary description of the kinematics and implementation of its actuators and sensors exposed in the second section, the third section models the TORVEastro arm control system according to the existence and uniqueness solution resolved in [3], describing the algorithmic approach. In the fourth and fifth sections, both the actuation control and experimental results are exposed.

© The Author(s), under exclusive license to Springer Nature Switzerland AG 2023
M. Ceccarelli et al. (Eds.): TORVEASTRO 2023, MMS 130, pp. 75–89, 2023.
https://doi.org/10.1007/978-3-031-28447-2_6

2 Robot Arm Kinematics and Its Architecture

As depicted in Fig. 1, the TORVEastro robot consists of a cylindrical central body [4], on which three identical, symmetrically arranged arms are mounted; each of the three curved arms, each consisting of three rigid mechanical links, are articulated by a rotating joint; this configuration gives the arm three degrees of freedom and makes it possible to use a gripper as end-effector. In addition, the curved design of each arm gives the robot more compactness when stored.

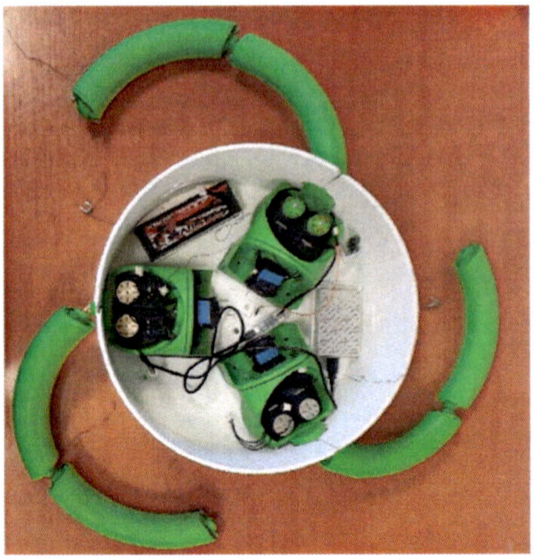

Fig. 1 Mechanical structure of the arms and internal layout

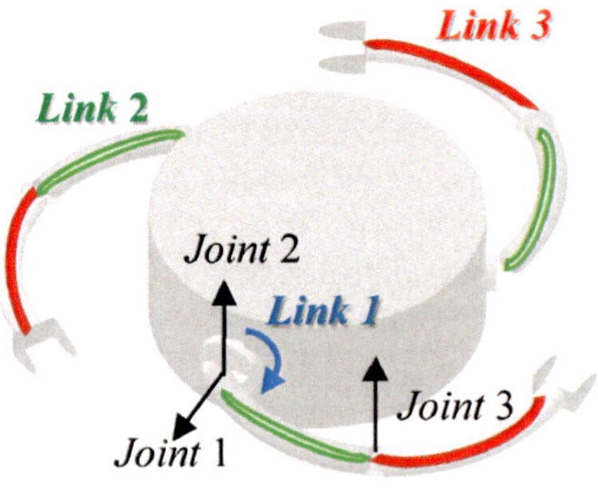

Fig. 2 Sketch of the TORVEastro robot

As discussed in [3], each motor unit inside the body (consisting of a single group of 3 motors per arm—Fig. 1) represents TORVEastro's anthropomorphic arm shoulder to perform the three DoF (Degrees of Freedom) of the arm. The motor unit consists of two components. The first component is connected to the body frame and contains an actuator (named motor number 1); this actuator is directly connected to the second component, allowing it to rotate 360 degrees. The second portion of the motor unit contains two actuators: one allows the cable-driven arm to move (named motor 2), while the other (named motor 3) allows the cable-driven forearm to move. Each arm and forearm are moved by two cables each.

Both actuators move the two pulleys (green and red one in Fig. 7) by performing flexion-extension movements directly related to the rotational direction.

The cables are routed through the arm and forearm to protect them from the environment.

As shown in Fig. 2, from a kinematic point of view, each TORVEastro arm constitutes a serial open kinematic chain, in which the first link is connected to the central body via the first joint (*joint 1*); the second link is connected to the first one via the second joint (*joint 2*) and the third link is connected to the second one via the third joint (*joint 3*). The second and third joints represents two passive axes of revolution, while the first joint is rigidly connected to its actuator. All nine actuators are arranged inside the body of the robot (Fig. 1) and the movements of the second and third joint for each arm are ensured by specific belts that connect the coaxial pulley of the corresponding actuator to its extreme destination of the controlled arm [5], running inside cavities, curved like the shaft of the arm. Thus, the extreme destination of the belt within the curved shaft forearm (*link 3*), passing through *link 2*, is the wrist, while the destination of the curved shaft arm belt is the elbow (*link 2*). Note that *link 1* corresponds to *joint 1*, so it can be called an active axis, otherwise the other joints, moved by their belt can be called passive axis. Therefore joint angles ϑ_{jk} describing the rotations around the joint *j* axis of the *k*-arm are adopted as the joint coordinates; the links reference frame was expressly assigned to define a joint angle as the angle from the $x_{(j-1)k}$ axis to the x_{jk} axis, which is considered positive for the counterclockwise direction around the $z_{(j-1)k}$ axis (see Fig. 3 and [3]).

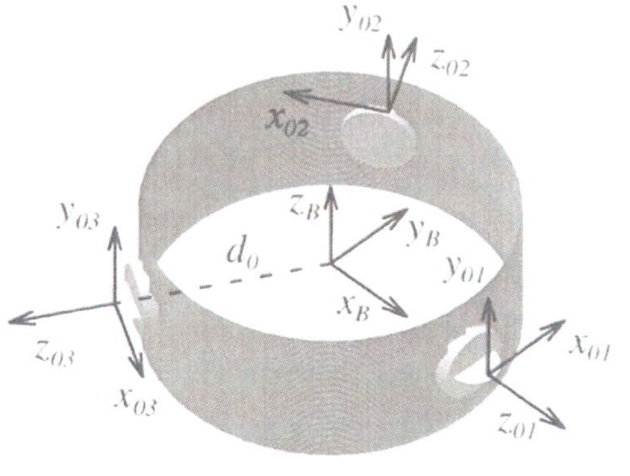

Fig. 3 Auxiliary body reference frames

The first joint allows the first link to rotate around an axis perpendicular to the lateral surface of the body (ϑ_{1k}), the second joint allows the second link to rotate around an axis that perpendicularity intersects the first joint (ϑ_{2k}) and the third joint allows the third link to rotate around an axis parallel to that of the second joint (ϑ_{3k}). During the movements of the arms, the orthogonality of the axes of the first and second joints and the parallelism of the second and third joints are always preserved. At the end of the chain, the end-effector is rigidly connected to the last link without any other joint.

During the arms movements, the positions and orientations of the auxiliary frames remain fixed in the body reference frame (Fig. 4), the common distance between their origins and the origin of the main body frame is denoted by d_O.

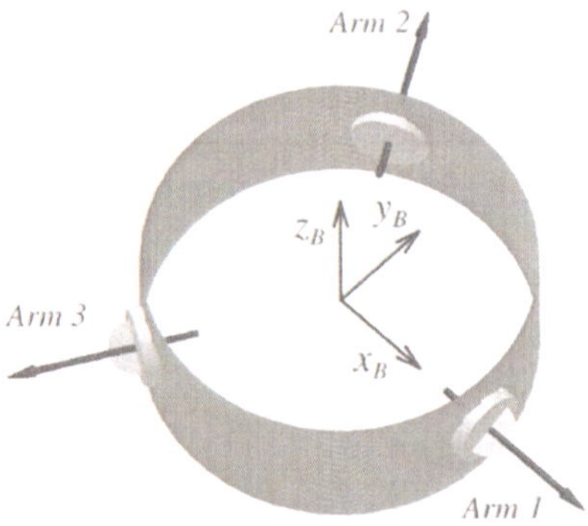

Fig. 4 Body reference frame

Each motor moves its respective link always knowing its current position, internal temperature and current draw, thanks to specific built-in sensors (e.g. a 12 bit encoder for position readout).

As in [3], safety acceptance limits were defined; in particular safe values for the angle ϑ_{1k} can range from $-180°$ to $180°$, for the angle ϑ_{2k} from $-150°$ to $30°$ and for the sum $\vartheta_{2k} + \vartheta_{3k}$ from $-120°$ to $60°$. However, in the real arm, described in this paper, the angles result in a different range, determined experimentally below.

Representing for the generic arm k:

$$\mathbf{A}_k(\vartheta_{1k}, \vartheta_{2k}, \vartheta_{3k}) = \mathbf{N_M} \cdot \vartheta_k + \mathbf{A}_k^0$$
$$\text{with} \quad \mathbf{A}_k = \begin{bmatrix} \alpha_{1k} \\ \alpha_{2k} \\ \alpha_{3k} \end{bmatrix}, \ \vartheta_k = \begin{bmatrix} \vartheta_{1k} \\ \vartheta_{2k} \\ \vartheta_{3k} \end{bmatrix} \text{ and } \mathbf{A}_k^0 = \begin{bmatrix} \alpha_{1k}^0 \\ \alpha_{2k}^0 \\ \alpha_{3k}^0 \end{bmatrix} \tag{1}$$

And by placing $\mathbf{A}_k$ the position vector assumed by all three motors for the arm k and $\mathbf{N_M}$ the scaling factor for the 3x3 matrix as in Eq. (1), the same for each arm; this formula allows us to convert ϑ_k (vector representing the three joint angles of a typical arm) into equivalent revolutions of the pulley that insists on the gear of all the three motors M of arm k (with k between 1 and 3). Note that, for coupling *joint 1*, no pulley is needed, since *link 1* corresponds to the rotation of the motor shaft itself (so called active axes).

$\mathbf{A}_k^0$ is a vector containing the initial angular position of each motor shaft/pulley.

$\mathbf{N_M}$ is determined experimentally by a form like this:

$$\mathbf{N_M} = \begin{bmatrix} R_{11} & 0 & 0 \\ 0 & R_{22} & R_{23} \\ 0 & R_{32} & R_{33} \end{bmatrix} \tag{2}$$

As mentioned, the rotations of the motor shaft around the axis of joint j of the arm k are adopted as the coordinates of the joint with ϑ_{jk}: to reach these values, it's seen experimentally that the motors pulley (numbered as the name of its joint) must vary approximately and respectively from $-180°$ to $180°$ for α_{1k} ($R_{11} = 1$), from about $60°$ to $115°$ for α_{2k} ($R_{22}\approx 2$) and from about $70°$ to $150°$ for α_{3k} ($R_{33}\approx 2.7$). Extra-diagonal elements can be chosen approximately as the average value between R_{22} and R_{33} ($R_{23}\approx R_{32}\approx 2.35$).

The results are depicted in Fig. 7 of Sect. 5.

In general, for the robot as a whole:

$$\mathbf{A}\left(\vartheta_{jk}\right) = \begin{bmatrix} \mathbf{A}_1 \\ \mathbf{A}_2 \\ \mathbf{A}_3 \end{bmatrix} = \mathbf{N_{Md}} \cdot \begin{bmatrix} \vartheta_1 \\ \vartheta_2 \\ \vartheta_3 \end{bmatrix} + \begin{bmatrix} \mathbf{A}_1^0 \\ \mathbf{A}_2^0 \\ \mathbf{A}_3^0 \end{bmatrix} \tag{3}$$

where $\mathbf{N_{Md}}$ is a 9x9 diagonal matrix with three identical $\mathbf{N_M}$ (3x3) matrices replicated on each diagonal element.

Note that the **A** (1x9) matrix is the position vector that passes through the built-in PID (Proportional-Integral-Derivative) control algorithm for each of the nine motors.

3 Modelling of the Process Control System

The motors chosen to move the robot arms, have the main features reported in Table 1 and each motor is factory-equipped with specific electronics acting on a local PID control. This allows autonomous management of each shaft respectively, in accordance with the control mode selected in the main algorithm control software, discussed in Sect. 4.

Table. 1 Common features of the actuators employed

Item	Specifications
MCU	ARM CORTEX-M3 (72 [MHz], 32Bit)
Motor type	Coreless
Position sensor	Contactless absolute encoder (12Bit, 360°)
Control algorithm	Embedded PID control
Resolution	4096 [pulse/rev]
Baudrate (TTL)	Up to 4,5Mbps
Backlash	20 [arcmin] (0.33°)
Gear ration	200:1
No load speed	63 [rev/min] @12 V
Operating voltage	12 V
Stall torque	~6 Nm @12V – 4,1A

In (Fig. 5) one can see the control process for each motor described by logical blocks referring to Position Control Mode, i.e. the drive mode chosen to act on the control.

This is one of six possible modes of local operation control; the position control mode can be summarized in a few steps up to pulley drive [6]: each motor, via its common bus shared with the other eight motors, stores its target position at a specific address, according to the desired position and velocity profile.

How the individual motors reach their position-target is derived from the internal Profile Velocity and Profile Acceleration memory field of each electronic motor, respectively, filled by the algorithm implemented and described in Sect. 4; the internal feed-forward and PID controller does the rest, calculating the PWM output for each desired trajectory and applying it to the motor through an inverter; finally, thanks to the on board motor sensors (encoders and current sensors), the internal electronics stores the actuation results (current position, speed, PWM and current) at specific addresses and returns then to the control software as inputs (see Sect. 4), to provide optimally driving of all nine robot motors.

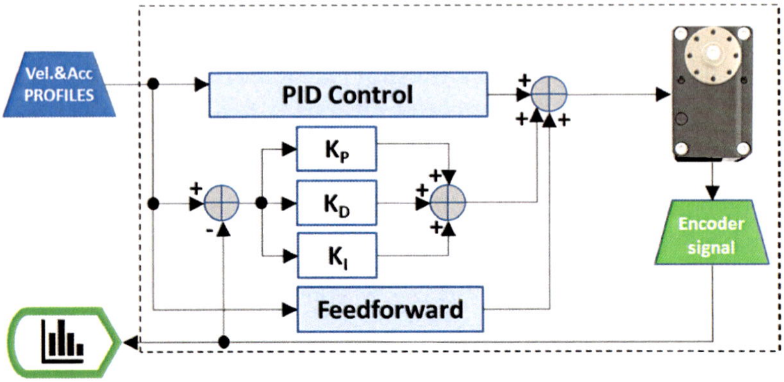

Fig. 5 Block diagram of the motor embedded PID for Position Control Mode

The Velocity Profile of the motor *k* for each arm emerges from the calculation based on both the ratio of the change in position angle (starting position minus new position to be reached) and the total time desired to reach the new position, which is identical for each motor, while the Acceleration Profile is derived from Velocity Profile to make the start and stop of the single movement smooth (set to zero for very low speed).

It was chosen that the local algorithm on board of each motor can act as a Time-Based Profile limited to a maximum of 32 s. If the total time desired for each trajectory exceeds this 32 s limit, the target position can be sliced into intermediate targets that extend the execution time, but with no guarantee on the total execution time. This limit is due to the minimum speed threshold of each motor.

To solve this problem, a preliminary time simulation procedure was developed in the main program to make each motor reaches its final position at the same time, always passing all motors a velocity above this lowest limit.

4 Actuation Control

According to [7], the central rule for optimal system control concerns the availability of on-board sensing and performance characteristics of the chosen motor, the latter being strictly related to its mechanical specifications (see Table 1).

The main requirements to meet them can be summarized as: embedded PID electronic control, metal gears–for a higher robustness of torque transmission–, reduced backlash–for a greater accuracy –, embedded processor and higher baudrate–for higher simultaneity of motion between motors–and, finally, TTL bus–for better compatibility of the communication standard with most popular control boards.

Assuming that an optimal control algorithm for a nine-motors system can be represented by the flow-chart depicted in Fig. 6.

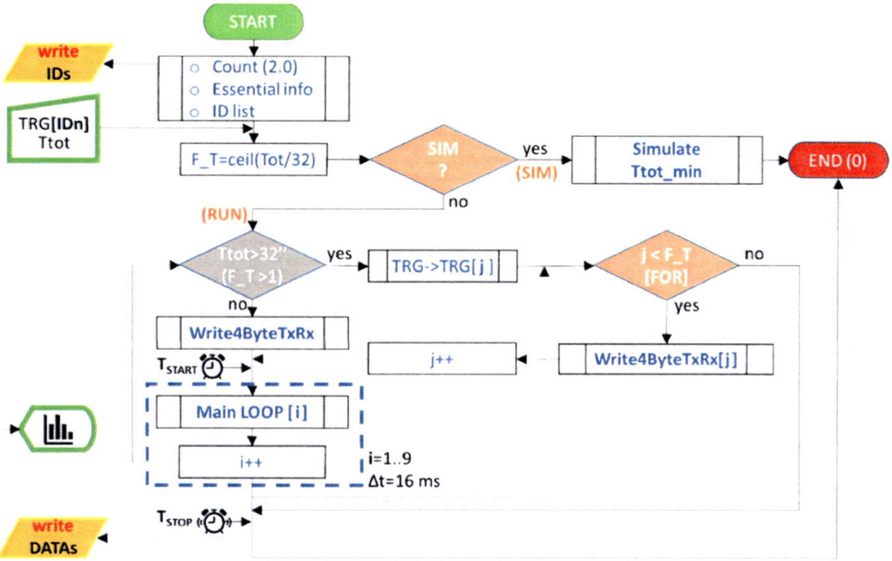

Fig. 6 Main program flow-chart for TORVEastro robot arms. Recursive *Main LOOP* procedure was bounded by dashed-line

The control program of the TORVEastro robot consists of two parts: a main program part (Fig. 6) and an internal recursive part (a procedure called *Main LOOP* in the same Fig. 6). The main program part has the peculiarity that is executed only once per trajectory, after an initial general quick scan of the motors wired to the Raspberry PI 4 control board [8] via USB to TTL adapter-compatible link working through 1Mbps multidrop TTL bus port.

Via the API (Application Programming Interface) [10] it was possible to identify each motor with its own ID address, from 1 to 9, by calculating the variable *F_T* (stands for *Fractional_Trajectory*) as a rounded fractional expression of the total execution time (represented by variable *Ttot* in this code), normally equal to or less than 1.

The program can be started either in simulation mode (SIM) or in execution mode (RUN). SIM mode calculates the minimum common time *Ttot* (see Sect. 3) as the time *Ttot* associated to the slowest motor, while RUN mode executes trajectories in *Ttot* time.

To minimize battery consumption, note that this total time *Ttot* is assumed to be the total execution time of the overall movement of the three arms of the robot, actuated simultaneously for all three trajectories.

Thus, under RUN conditions, this first part of the program also acts, after the recursive part inside it, on both the counting of the final real elapsed time for the execution of the overall trajectories and the process of saving data for all motors.

If the requires *Ttot* exceeds 32 s, the main program—before RUN—splits target the α_{jk} into a certain number of sub-trajectories and their respective target steps (*TRG[j]* in Fig. 6), based on the variable *F_T*. Reaching the final target or intermediates target positions are carried out by the so called *Main LOOP* block in Fig. 6, which represents the recursive part of the program. This is based on API interfaces [10] both to recursively read at specific addresses from each motor and to write all read values to the control board unit memory; it is executed until the individual position of each motor approaches the target value within a certain tolerance (set at about 0,4°, about 5 times the encoder resolution).

The final real elapsed time will be the difference between T_{STOP} and T_{START}, as shown in Fig. 6.

This condition is checked at each cycle, simultaneously for each motor, in an estimated time per iteration of about 16ms.

Note the particular case of $F\_T = 1$: the final target and the last intermediate target match and right-part of the flow-chart in RUN mode (FOR condition in Fig. 6) is not executed.

5 Experimental Results

A preliminary experimental-test was conducted in [3] on the rotating section of the robot only, connecting the motor shaft of *joint 1* with the robot shoulder (*link 1* in Fig. 2) but without any arm mounted and controlling only when the target position was reached. The present work complements the experimental proof treated in [3] for the remaining two joints by extending the possibility of controlling the whole robot by knowing the solution matrix $\mathbf{A}\left(\vartheta_{jk}\right)$ mentioned in (3) and according to the algorithm exposed in the previous section.

In this work note that it was also not possible to replicate the rotational motion as in [3], since the robustness of the embedded joints was inadequate to perform this kind of test; however a new type of joint was submitted for patent release [11].

Therefore the angle of *joint 1* was fixed at 90°, leaving the arm free to move on a flat surface, lying on a horizonal plane.

The test-bed of the generic arm is depicted in Fig. 7. Specifically, in Fig. 7a the extreme angles covered by the green pulley (identified by number 2) and the red pulley (identified by number 3) are represented by green and red vectors, while semi-transparent arcs respectively highlight all the possible trajectories for both individual links (the green arc indicates all the possible movement of the arm, while the red arcs indicate that of the forearm); also in Fig. 7a all possible arm positions are represented (the light blue arc traces the movement of the arm with the forearm always completely extended, while orange arc traces the movement of the arm with the forearm always completely constrained). Finally, in Fig. 7b, we can see the position of all possible points lying in a plane reachable by the arm wrist.

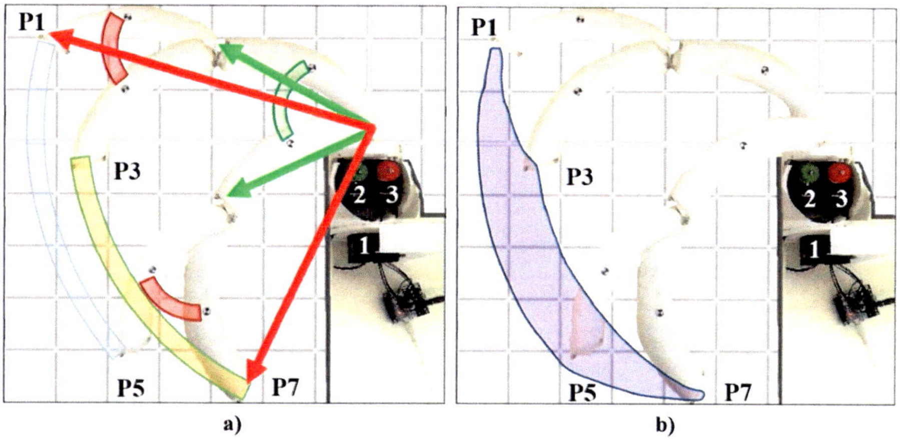

Fig. 7 **a)** Absolute position representation (P1-P3-P5-P7) and **b)** all possible reachable points by the wrist of the arm, lying in a plane, *joint 1* fixed at 90°

As a result of this experimental outcomes, it is possible to consider the decomposition of all the complex robot movements for all nine actuators into most simpler trajectories, such as those depicted in Fig. 7, as a sequence of single arm movements, provided that the purple area in Fig. 7b concerning reachable points is respected and in accordance with the algorithm depicted in Fig. 6 (see Sect. 4).

All x-axis units are expressed in a dimensionless control-cycle counter (progressive number of iterations in the *Main LOOP* code), so a multiplication by 0,016 returns the value in seconds.

All simple fundamental movements (also called modes) are therefore identified in Table 2, ranging from P1 (extreme starting-point position, also called starting position in [9]) to P7 (extreme end-point position, also called ending position in [9]), passing through P3 and P5.

The experimental results for a single generic arm of the TORVEastro robot are shown. In particular, from Figs. 8, 9, 10 and 11, all angular changes from the starting position to the target position are shown; the units of the y-axes are in degrees.

From Figs. 12, 13, 14 and 15, instead, all the respective angular velocities achieved in the same motion are shown; the units of the y-axes are in rpm.

All intermediate movements from a generic **X** position (Fig. 7b) are possible through the intermediate positions, according to the general formula given in Eq. (3) (see Sect. 2), posing, for example in Fig. 8, $\mathbf{A}_k^0 = \mathbf{A}_k^{P1}$, , according to (1).

It should be noted that a possible facilitation to reach the final position relative to the movement M5 should be achieved by moving both *links 2* and *3* (both arm and forearm)

Table. 2 Simple fundamental movements (modes) of a generic arm.

Fundamental movement name (modes, from start position to end position) **X**: generic position	Part moving to **C** or to **O** (and direction of rotation) **C**: Close – **O**: Open **CW**: ClockWise **CCW**: CounterClockWise	Pulley Relative Rotation (α_{jk}- α_{jk}^{0}) [°] **x**: generic angle
M1 (**X** to P1)	Arm **O** (CCW) + Forearm **O** (CCW)	-x / -x
M3 (P1 to P3)	Arm **O** + Forearm to **C** (CW)	**0 / +110**
M5 (P3 to P5)	Arm **C** (CW) + Forearm to **O** (CCW)	**+110 / 0**
M7 (P5 to P7)	Arm **C** + Forearm to **C** (CW)	**0 / +115**

first to P7 and then to P5, to reduce friction in the cavity arm module during the closing process.

Finally, the images from Figs. 16, 17 and 18 represent a single execution of the algorithm to bring the single arm (named arm 1) from the position P7 to P1 in 15 s; these plots refer to the starting point at 5.32 s.

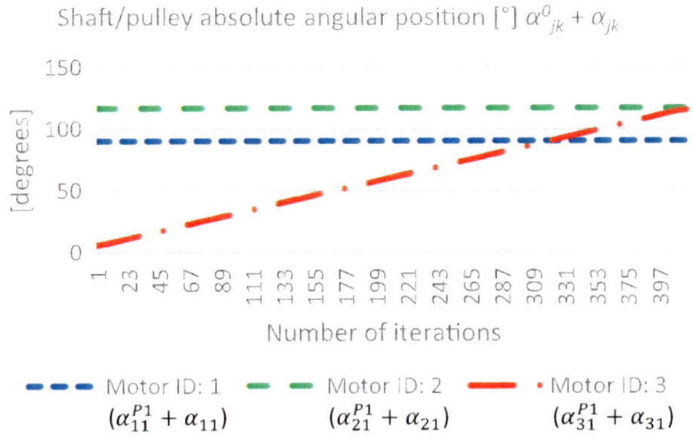

Fig. 8 Pulley position angle of a trajectory transition from P1 to P3 (M3 movement)

Note in Fig. 17 the slight deviation from zero manifested by *link 2* (arm), observed between seconds 5.32 and 10, which, while appearing to be almost stationary to con-sent the motion of *link 3* (forearm) alone, actually goes along with the overall motion of the arm (Fig. 18) with regularity from the beginning of execution (second 5.32), precisely to allow it to fully reach position P1.

Additionally, the use of nylon belts for the transmission of motion over the entire arm was found to be necessary to facilitate sliding in the cavities, but this introduced

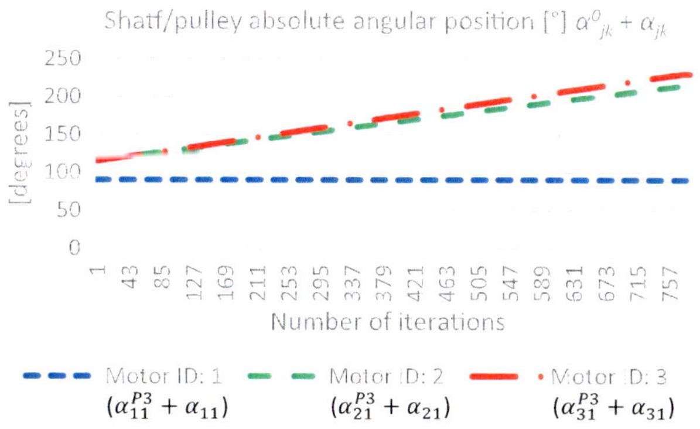

Fig. 9 Pulley position angle of a trajectory transition from P3 to P7 (M7 movement)

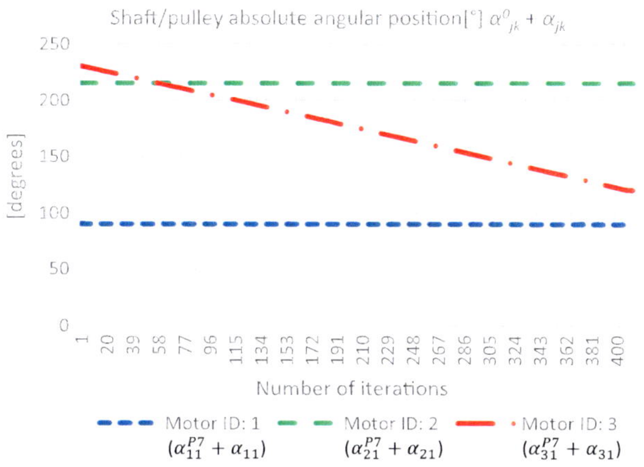

Fig. 10 Pulley position angle of a trajectory transition from P7 to P5 (M5 movement)

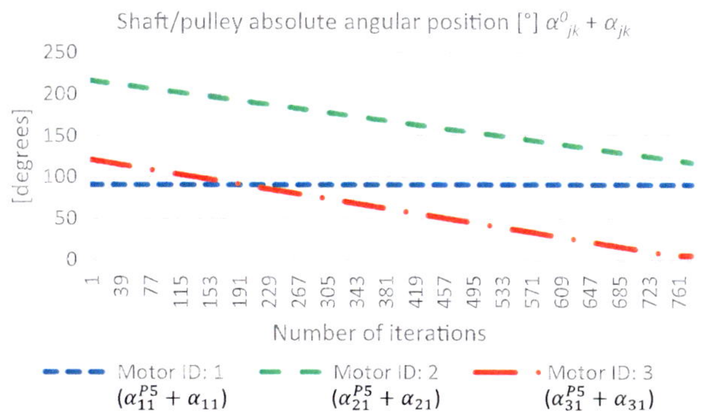

Fig. 11 Pulley position angle of a trajectory transition from P5 to P1 (M1 movement)

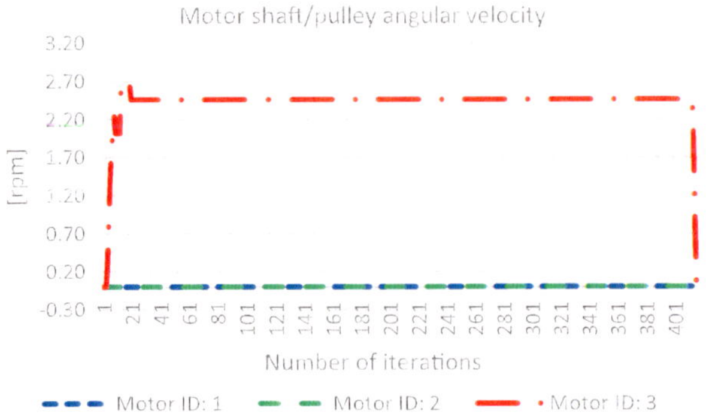

Fig. 12 Trajectory velocity of a transition from P1 to P3 (M3 movement)

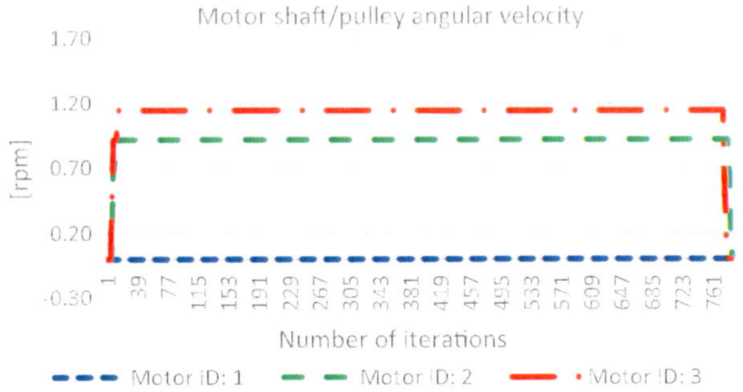

Fig. 13 Trajectory velocity of a transition from P3 to P7 (M7 movement)

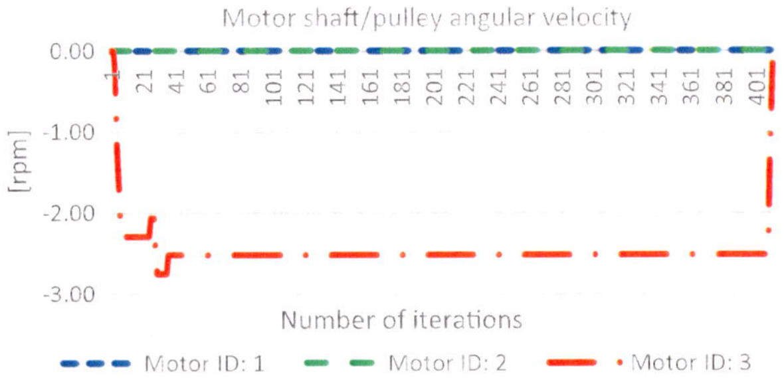

Fig. 14 Trajectory velocity of a transition from P7 to P5 (M5 movement)

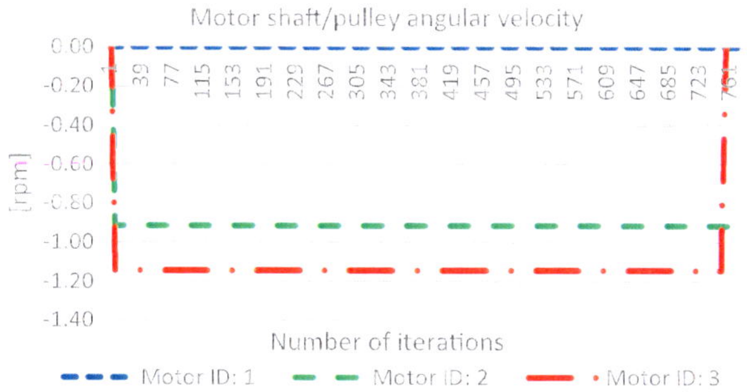

Fig. 15 Trajectory velocity of a transition from P5 to P1 (M1 movement)

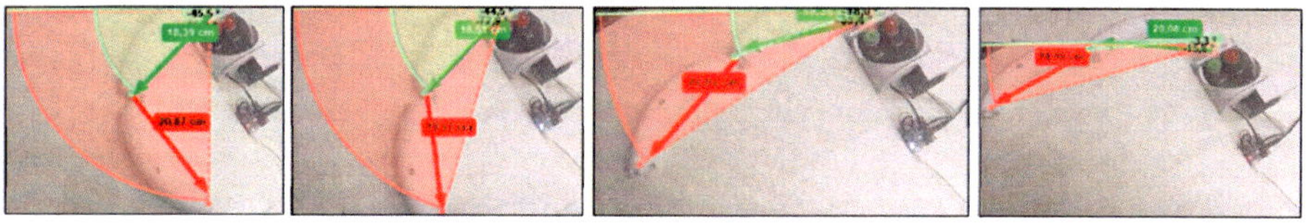

Fig. 16 Screenshots of an arm trajectory tracing during the transition from P7 to P1

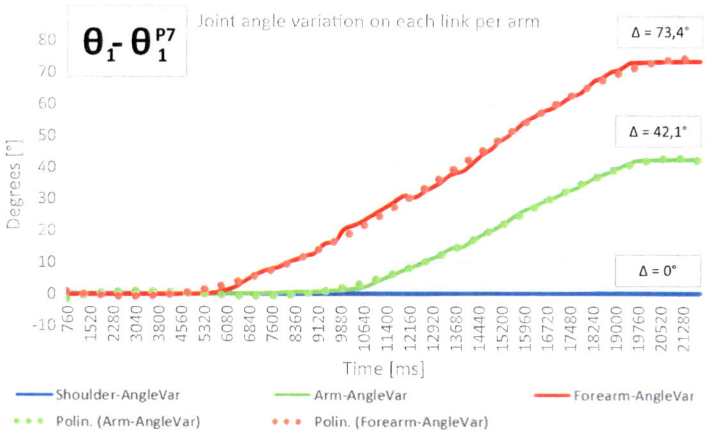

Fig. 17 Final graph of overall single-arm trajectories from P7 to P1

additional relaxations to the stresses predicted at the design stage due to the known elasticity of the material chosen for the transmission of force.

6 Conclusions

The control design of the TORVEastro prototype is described. As a result of this experiment outcomes, it can be concluded that a preliminary test-bed was set up for a single arm of TORVEastro and an initial characterization of the single arm of the robot could be performed. All outcomes are aligned with both the joint configuration chosen during

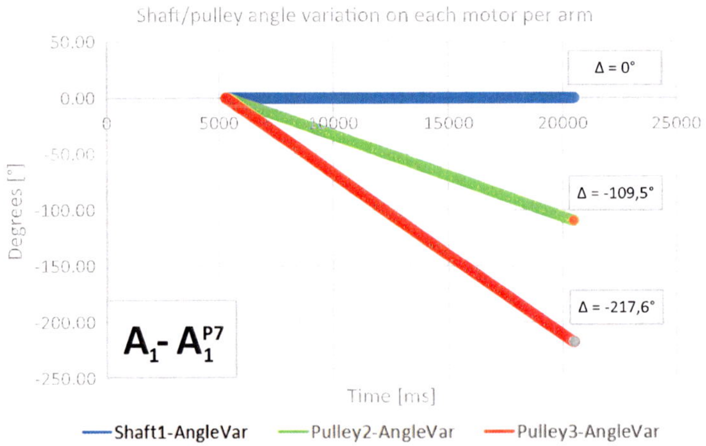

Fig. 18 Final graph of overall single-arm shaft/pulley angles variations from P7 to P1

the execution of the algorithm and the expected time to reach them. However, some limitations are highlighted below.

Due to the internal limiters related to the PID electronics embedded into the motor, it appears very difficult, under certain conditions, to accurately control the time to reach the final destination (in short, the total real elapsed time T_{STOP} obtained at the end of *Main LOOP*–see Fig. 6–might be greater than the total required execution time).

A possible future improvement would be to extrapolate the initial slope of the slowest motor's position (experimentally observed to be constant throughout the entire movement) and readjust the speed and acceleration/deceleration profiles of all other faster motors to the predicted and estimated total final execution time of the slowest motor.

Driving the arm with nylon cables also brought significant deviations from the expected reduction ratios alluded to in Sect. 2. However, those experiments allowed, net of the cable that will be used of the final demonstrator, to fine-tune the algorithm that will be called upon to govern, at the lowest control level, the entire robot.

Finally, as an additional future improvement, to better execute movements according to M5 (see Table 2), an exchange of belts with respect to their cavity on the arm module pulley (number 2 in Fig. 7) can be evaluated to reduce friction as much as possible.

Acknowledgments. The TORVEastro project activities reported in this paper were funded by Regione Lazio [Project 36517 POR FESR LAZIO 2014–20].

References

1. Samani, F., Ceccarelli, M.: Prototype Design and Testing of TORVEastro, Cable-Driven Astronaut Robot. In: Zeghloul, S., Laribi, M.A., Sandoval Arevalo, J.S. (eds.) RAAD 2020. MMS, vol. 84, pp. 448–455. Springer, Cham (2020). https://doi.org/10.1007/978-3-030-48989-2_48

2. Araque-Isidro, J.E., Cafolla, D., Ceccarelli, M.: Problems and requirements for outer space astronaut service robot. In: Niola, V., Gasparetto, A., Quaglia, G., Carbone, G. (eds.) Advances in Italian Mechanism Science IFToMM Italy 2022, pp. 603–611. Springer, Cham (2022). https://doi.org/10.1007/978-3-031-10776-4_69
3. Paoloni M., Cupertino, G., Ceccarelli M.: A kinematic analysis of the TORVEastro astronaut robot arm. In: 13th IFToMM International Symposium on Science of Mechanisms and Machines (SYROM 2022), November 17–18, Iasi, Romania, Paper No. 28, Springer, Cham (2022) (in print)
4. Cafolla, D.: Design and preliminary testing of Torveastro, an outer space service robot. In: Proceedings of the 13th IFToMM Inte. Symposium on Science of Mechanisms and Machines (SYROM 2022), November 17–18, Iasi, Romania, Springer, Cham (2022) (in print)
5. Ceccarelli, M., Qiang, H., Hui, L.: Robot astronauta con tre arti. Italian patent number 102018000007507, application filed on July 25, 2018 (2020)
6. Robotis.: MX-64-T/2.0 Manual. https://emanual.robotis.com/docs/en/dxl/mx/mx-64-2. Accessed 22 Nov 2022
7. Ceccarelli, M.: Fundamentals of Mechanics of Robotic Manipulation, 2nd edn. (Series: Mechanisms and Machine Science). Springer (2022)
8. Raspberry pi documentation. Raspberry Pi hardware. https://www.raspberrypi.com/documentation/computers/raspberry-pi.html Accessed 22 Nov 2022
9. Cafolla, D.: Performance analysis of Torveastro, an outer space service robot. In: Proceedings of TORVEASTRO 2023 conference, 20–21 April 2023, Rome, paper 04, Springer, Cham (2023) (in print)
10. Robotis.: Dynamixel SDK Manual. https://emanual.robotis.com/docs/en/software/dynamixel/dynamixel_sdk/api_reference/c/c_porthandler Accessed 22 Nov 2022
11. Ceccarelli, M.: TORVEASTRO project: plans and achievements. In: Proceedings of TORVEASTRO 2023 conference, 20–21 April 2023, Rome, paper 01, Springer, Cham (2023) (in print)

Check for updates

Performance Analysis of Torveastro, an Outer Space Service Robot

Daniele Cafolla(✉)

LARM-Laboratory of Robot Mechatronics, University of Rome Tor Vergata, Rome, Italy
contact@danielecafolla.eu

Abstract. For on-orbit servicing (OOS) tasks including as docking, berthing, refuelling, repairing, upgrading, transporting, rescuing, and orbital trash removal, space robots is considered as one of the most promising options. In the previous two decades, numerous enabling techniques and technological demonstration missions have been created and executed. Several manned on-orbit service missions have been completed successfully, but entirely unmanned servicing missions have not yet been conducted. Robotic servicing is still an active topic of research with numerous technical obstacles. This paper describes the design and preliminary testing of Torveastro, a service robot for outer space. The specifications are supplied alongside the design and simulation. Finally, simulation results showed a promising behaviour for further developments of the prototype.

Keywords: Space robotics · Service robotics · Robot design · Performance analysis · Multibody simulation

1 Introduction

In the past decade, an average of one hundred satellites were launched annually, according to statistical data. Most of them accomplished their missions without significant obstacles. Nevertheless, a small percentage of them experienced anomalies and even failures of varying severity [1]. Launcher failure was the most common cause of failure in the past. However, on-orbit failures have surpassed launch failures for the first time in recent years [2] and cumulatively account for billions of dollars in losses [3]. Moreover, every launched satellite eventually runs out of fuel, necessitating its decommissioning even if it is still operational [4]. Several studies [5–7] have demonstrated the cost effectiveness of in-flight repair of damaged spacecraft, and Sale et al. [8, 9] have presented a model that incorporates risk and uncertainty analysis. In order to protect their assets in space, the National Aeronautics and Space Administration (NASA) realized as early as the 1980s [10] the significance of robotics in orbit servicing operations. After their deployment, on-orbit servicing (OOS) refers to the maintenance of space systems in orbit, including the repair, assembly, refueling, and/or upgrade of spacecraft. Notably, such complex space missions have prompted the development of new space robotics technologies and a number of experimental demonstration missions, including both manned and unmanned missions [10]. A typical space robotic system (also known

© The Author(s), under exclusive license to Springer Nature Switzerland AG 2023
M. Ceccarelli et al. (Eds.): TORVEASTRO 2023, MMS 130, pp. 90–99, 2023.
https://doi.org/10.1007/978-3-031-28447-2_7

as a space manipulator or space robot) for an OOS mission consists of the following three major components: a base spacecraft or servicing satellite, a robot manipulator with n degrees of freedom (n-DOF) attached to the servicing satellite, and the target spacecraft to be serviced. A vehicle for servicing spacecraft manipulators is sometimes referred to as the servicing system. Since the first deployment of the Shuttle Remote Manipulator System (SRMS) from the cargo bay of the Space Shuttle Columbia in 1981, robotic systems have been utilized on a variety of space missions, including the International Space Station (ISS) [11]. Developed by the Canadian Space Agency (CSA), SRMS is a 15.2 m long, six-degrees-of-freedom (DOF) robotic manipulator known as Canadarm, which has performed a number of on-orbit service missions [12].

As a result of the advancements in robotics technologies, some of the current space manipulators are able to move with great dexterity, enabling them to assist or even replace astronauts in performing precise, complex, or dangerous tasks. Several experimental space manipulators have been developed and successfully flown in space, for instance. Later, DLR, the national aeronautics and space research centre of the Federal Republic of Germany, developed the robotics experiment ROKVISS (Robotics Component Verification on the International Space Station) [13]. NASA and General Motors developed the second Robonaut, R2, for flight testing on the International Space Station. This advanced, dexterous, anthropomorphic robot has significant technical advancements over its predecessor, making it a significantly more valuable tool for astronauts. Increases in force sensing, range of motion, bandwidth, and dexterity are among the enhancements [14]. Robotic manipulators are ideally suited for repetitive tasks that would be too time-consuming, dangerous, and costly for astronauts to perform. In addition, space robots require significantly less infrastructure than humans (such as life support systems), making them preferable to humans in space. Consequently, this paper presents the design, simulation, and preliminary tests of Torveastro, a space service intelligent robot for the on-orbit operation requirements of space stations that is able to effectively utilize handrail-equipped installations designed for the movement of human astronauts during outer space operations.

2 Design

Microgravity acting on the robot, high radiation from the sun or other radiation sources, and sudden and extreme temperature changes are distinguishing features of the space environment. These environmental constraints can even cause actuators and electronic devices to malfunction [15].

In addition to autonomous tasks, the robot may be able to collaborate or assist astronauts with construction, maintenance, and monitoring. As a result, adequate robot mobility is essential, as it guarantees work outside of a space station. The robot may move slowly while maintaining consistent action results and a firm posture. Consequently, Torveastro is designed to move along the handrails intended for human space travel. The handrails are installed along the station's surface so astronauts can perform maintenance and service operations in outer space. The handrails have a compact structure of (34.95x15.9) mm each, are extremely stable and easy to grasp, and are separated by 1 m [16]. Figure 1 depicts the installation of handrails in space.

In [17], requirements and characteristics are reviewed to identify design issues and operational characteristics. Additionally, a dimensional design of Torveastro is presented, as well as the computation and modelling of its design parameters. Figure 2 depicts a schematic representation of Torveastro with the computed dimensional parameter that enables consistent action results and a stable posture while moving and performing outer space tasks with the installed handrails h described above. Torveastro has a cylindrical body with a diameter *b* of 370 mm, three arms comprised of two arc-shaped limbs *a* of 195 mm, resulting in a total arm length of 390 mm, and a distance between two extended arms that exceeds 1 m. Finally, In launching position, the arm can fully embrace the body due to its arc shape.

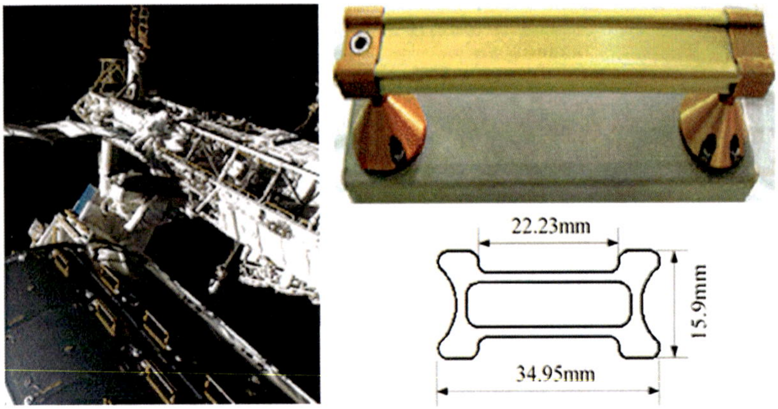

Fig. 1. Examples of handrails for astronaut robot motion detail, [16].

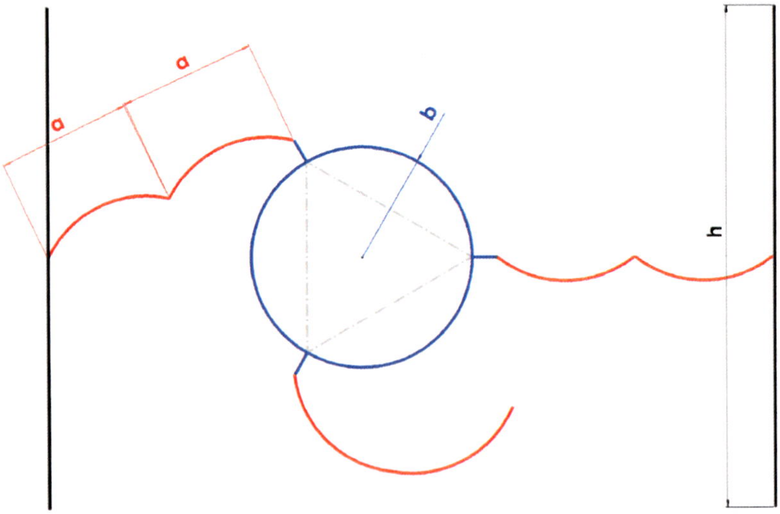

Fig. 2. Schematic representation of the prototype.

With the knowledge of the dimensional parameters, a 3D CAD has been implemented to detail the component of Torveastro that enables it to perform the desired movement while maintaining the actuation, control, and electronics within the body to protect them from the environmental conditions and making the design as compact as possible. Figure 3 shows the final CAD of Torveastro. The motor unit detail is depicted in Fig. 3a. The motor unit represents the shoulder of Torveastro's anthropomorphic arm, and it contains three actuators to perform the arm's three DoF (Degrees of Freedom). The motor unit consists of two components. The first component is attached to the body frame and contains an actuator; this actuator is directly attached to the second component, allowing it to rotate 360 degrees. The second portion of the Motor unit contains two actuators: one allows the cable-driven arm to move, while the other allows the cable-driven forearm to move. Both actuators move the two pulleys by performing flexion-extension movements directly related to the rotational direction. Figure 3b depicts the Full Arm's four components. The interface between the Motor unit and the Arm, the Forearm, and the End-Effector. The Motor unit interface permits the attachment of the Full Arm to the former. As previously stated, the arm and forearm are moved by two cables each. The cables are routed through the arm and forearm to protect them from the environment. Figure 3c, shows a section of the full arm where the cables guide through the forearm and the arm up two the body are visible. It can be noticed that the guides are two in the internal part and two in the external part taking the center of the limb as reference, each pair is needed two perform the flexion and extension of a single limb. The final portion of the body is a two-finger End-Effector that is powered by a separate motor. The designed prototype weighs approximately 50N and the full CAD assembly is shown in Fig. 3d.

The design of the Torveastro astronaut robot is centred on the features of a low-cost terrestrial demonstration that will demonstrate the characteristics of the proposed Torveastro architecture. The adopted kinematic must ensure the transmission of motion and forces to the joints of the limbs so that the robot can conduct manipulations and locomotion in the external structures of orbital stations with dexterity.

3 Motor Unit Characterization via Simulation

ABS material from Table 1 has been selected for the simulation and prototype of Torveastro. With the full CAD prototype and all its characteristics, an 8-s multibody simulation was performed to simulate a real-world operation scenario and evaluate the feasibility and behavior of the proposed solution. A half rotation 180 degrees of the shoulder forward and backward to the starting position applied as shown in Fig. 4a. In addition, a flexion of the arm was simulated by applying an angular displacement as depicted in Fig. 4b, and a reaction torque of 2.80 N was applied at the same point to simulate the full arm weight of 10 N carrying a payload of 90 N payload while pulling the cable. Furthermore, the full arm's 100 N payload applied to the motor unit's edge as shown in Fig. 4c. Finally, the friction between bodies and terrestrial gravity have been considered in the simulation.

Figure 5 depicts the angular displacement provided as an input for shoulder actuation, whereas Fig. 6 depicts the torque required to execute the given input motion. The motor

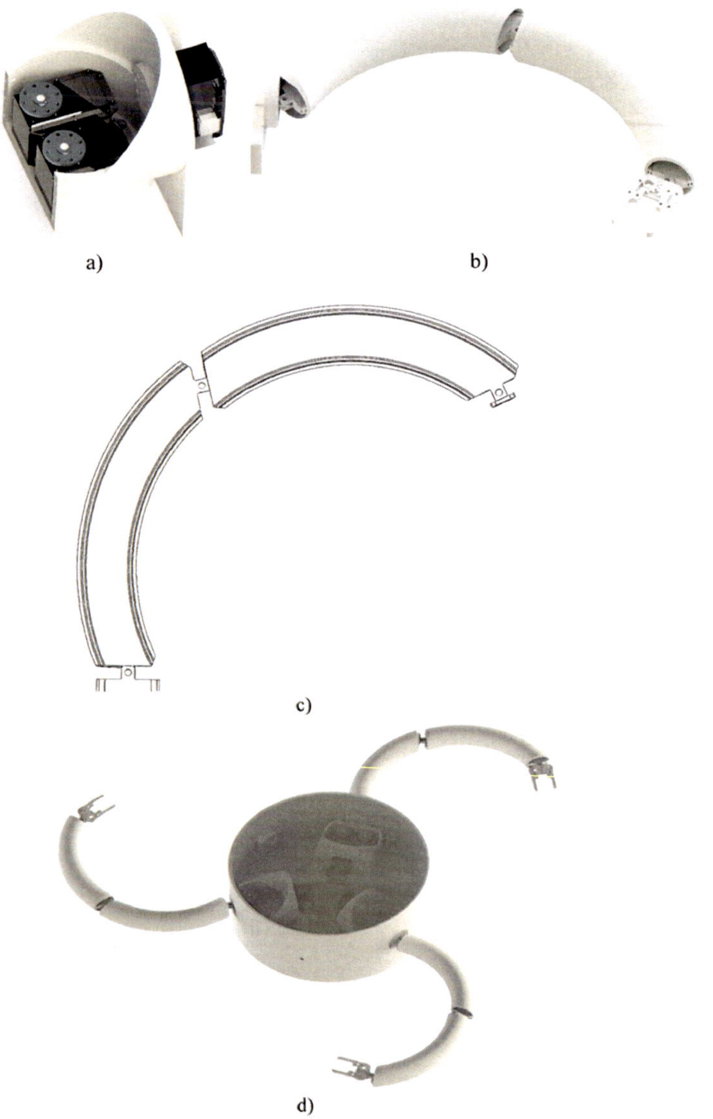

Fig. 3. Torveastro CAD: a) Motor Unit detail, b) Arm Detail, c) Cable guides detail section, d) Full assembly.(continue)

begins to move from its initial position of 0.00°, reaching 180.00° at second 4.00, and then returns to 0.00° between seconds 4.00 and 8.00. The torque depicted in Fig. 7 is clearly divided into two parts: the portion arising from second 0.00 to 4.00, where the torque reaches a maximum of 3.13 Nm at second 2.32. At second 4.00, the actuator of the shoulder joint reverses direction to return to the starting position, achieving a peak torque of 5.07 Nm at second 6.04. To contrast the terrestrial gravitational force and maintain the same motion profile, a greater torque is generated for the descending phase.

Table 1. ABS characteristics.

Elastic modulus	2.00 x 10^9	N/m^2
Mass density	1020	kg/m^3
Tensile strength	3.00 x 10^7	N/m^2
Thermal conductivity	0.23	W/(m·K)
Elastic modulus	2.00 x 10^9	N/m^2

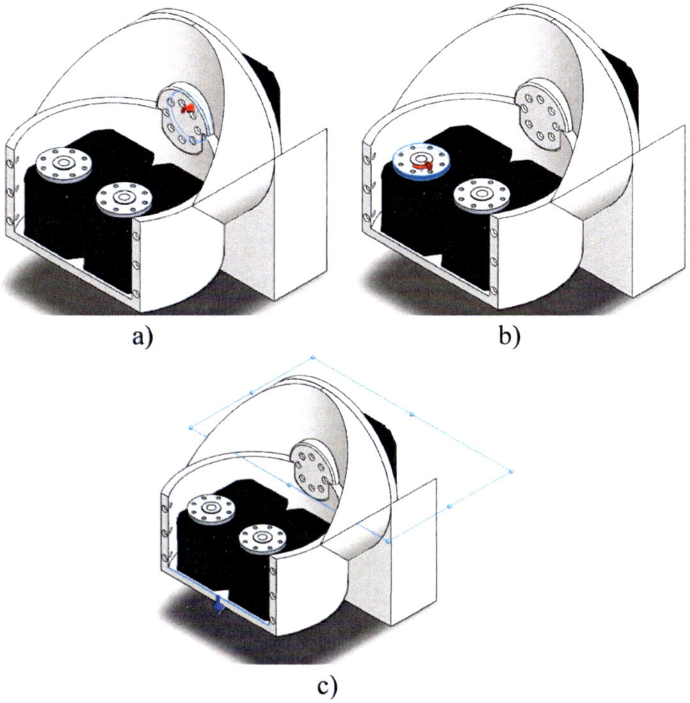

Fig. 4. Simulation models: a) Angular displacement of the shoulder application detail, b) Angular displacement of the arm application detail, c) Full arm payload force application detail.

At second 4.00, the motor that pulls the cable to simulate arm flexion is activated and encounters the aforementioned torque of resistance. In Fig. 8 it can be noticed that for the actuator to perform the desired operation of the arm, a maximum torque of 2.80 Nm was required at second 8.00. As seen in the graphs, the proposed solution has a smooth behavior and can be implemented using commercially available intelligent actuators.

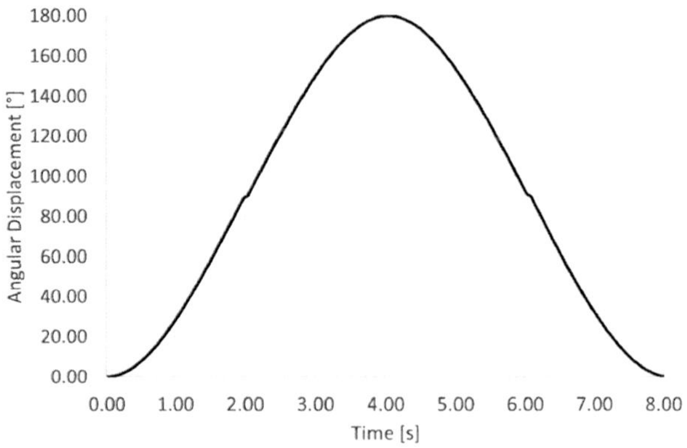

Fig. 5. Motor unit simulation input.

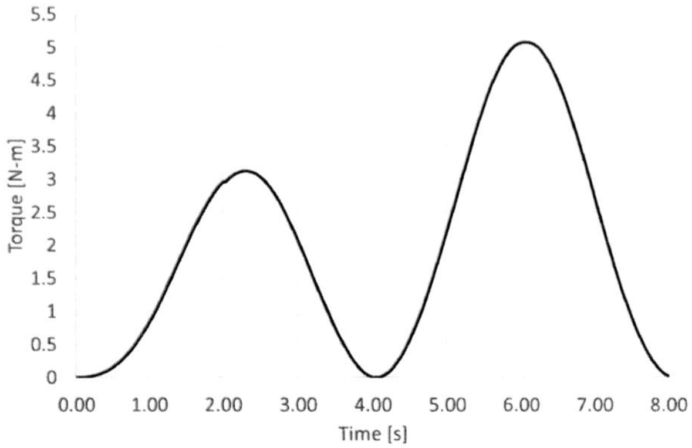

Fig. 6. Computed torque for motor unit shoulder motion in the motor unit simulation.

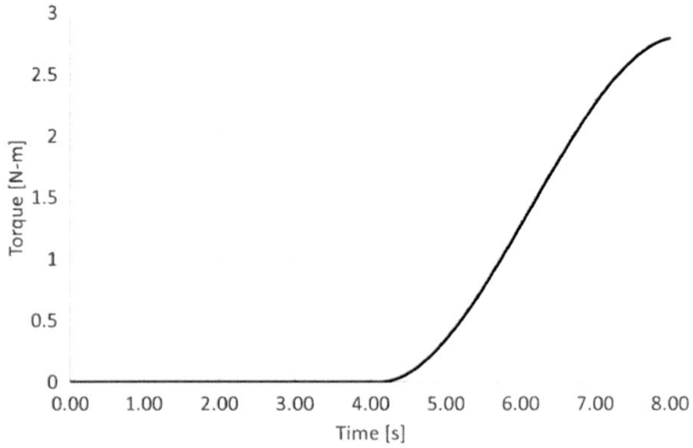

Fig. 7. Computed torque for arm actuator motion in the motor unit simulation.

4 Full Arm Characterization via Simulation

In the following phase, a 5 s displacement input was used to simulate the complete arm and have a better characterization. At the end effector, the identical modes and parameters stated previously have been applied to a 90N payload. As depicted in Fig. 8, an angular displacement input for the arm and forearm has been supplied for a deflection operation of 30 degrees and 60 degrees, respectively. The simulation models are represented in Fig. 9. Figure 9a shows the starting position of the simulation, while Fig. 9b shows the final position. Both in Figs. 9a and 4b the full trajectory of the arm and forearm can be noticed (Fig. 10).

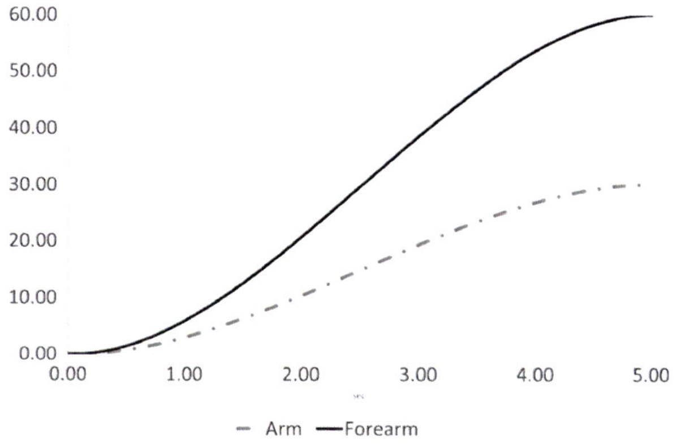

Fig. 8. Full arm simulation input.

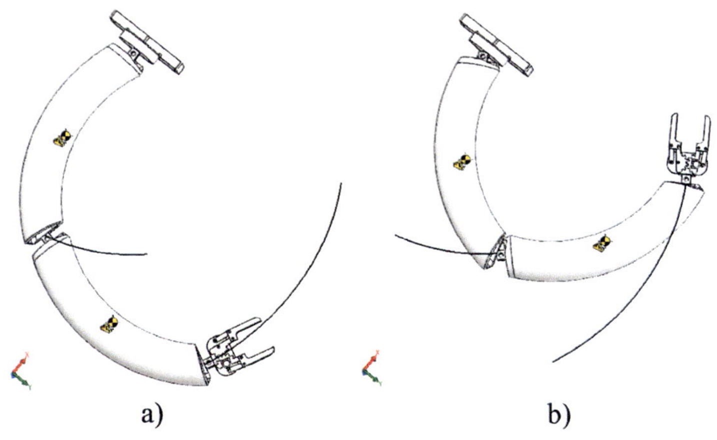

Fig. 9. Simulation modes for the full arm flexion: a) Starting position, b) Ending position.

Based on the results, the Dynamixel MX-64T [18] actuators were selected according to their qualities. The motor unit will be provided with a LiPo 14.8V 6200mAh battery in order for the actuator to attain the required performance. As a microcontroller, a Raspberry Pi 4 [19] will be coupled to a U2D2 Robotis interface [20].

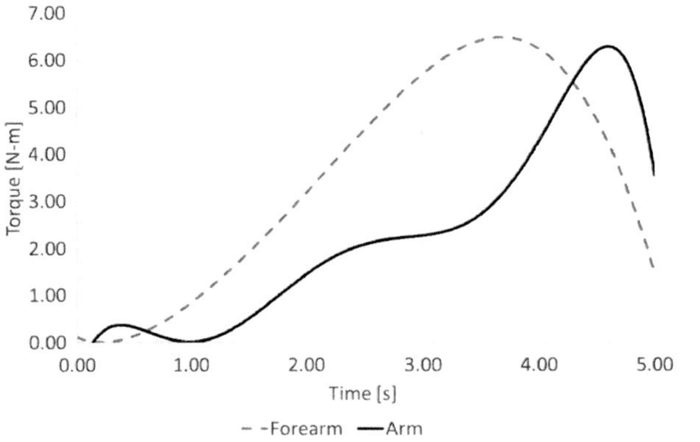

Fig. 10. Simulation results for the computed torques in the full arm simulation.

5 Conclusions

The design of the outer space service robot Torveastro is discussed. Using simulations, the feasibility of the proposal has been investigated. Simulating both the Motor Unit and the Full Arm allowed for a comprehensive characterisation of each. Simulations indicated a promising behavior for future advancements, enabling comprehension of the prototype's components. Future efforts will depend on a detailed full assembly and its manufacture and testing.

References

1. Sullivan, B., Akin, D.: A survey of serviceable spacecraft failures. In: AIAA Space Conference and Exposition, vol. 4540. Albuquerque, NM (2001)
2. Ellery, A., Kreisel, J., Sommer, B.: The case for robotic on-orbit servicing of spacecraft: spacecraft reliability is a myth. Acta Astronaut **63**(5), 632–648 (2008)
3. Landis, G., Bailey, S., Tischler, R.: Causes of power-related satellite failures. In: IEEE Fourth World Conference on Photovoltaic Energy, vol. 2, pp. 1943–5. Waikoloa, HI (2006)
4. Martin, D.: Communication Satellites. American Institute of Aeronautics & Astronautic (2000)
5. Waltz, D.M., Cepollina, F.: On-orbit Servicing of Space Systems. Krieger Publishing Company (1993)
6. Davinic, N., Chappie, S., Arkus, A., Greenberg, J.: Spacecraft modular architecture design study–cost benefit analysis of on-orbit satellite servicing. In: IAF—International Astronautical Congress, Turin, Italy (1997)
7. Leisman, G., Wallen, A., Kramer, S., Murdock, W.: Analysis and preliminary design of on-orbit servicing architectures for the GPS constellation. In: AIAA Conference and Exposition (1999)
8. Saleh, J.H., Lamassoure, E., Hastings, D.E.: Space systems flexibility provided by on-orbit servicing: part 1. J Spacecr Rockets **39**(4), 551–560 (2002)
9. Lamassoure, E., Saleh, J.H., Hastings, D.E.: Space systems flexibility provided by on-orbit servicing: part 2. J Spacecr Rockets **39**(4), 561–570 (2002)

10. NASA. On-orbit satellite servicing study project report. Technical Report, NASA; Zhang, T., Chen, Z., Wang, X., Liang, B.: Overview and prospect of key technologies of teleoperation of space robot. Aerosp. Control Appl. **40**, 1–9 (2014)
11. Flores-Abad, A., Ma, O., Pham, K., Ulrich, S.: A review of space robotics technologies for on-orbit servicing. Prog. Aerosp. Sci. **1**(68), 1–26 (2014)
12. Rembala, R., Ower, C.: Robotic assembly and maintenance of future space stations based on the ISS mission operations experience. Acta Astronaut. **65**(7), 912–920 (2009)
13. Hirzinger, G., Landzettel, K., Reintsema, D., Preusche, C., Albu-Schäffer, A., Rebele, B., et al.: ROKVISS—robotics component verification on ISS. In: Artificial Intelligence, Robotics and Automation in Space, San Diego, CA (2005)
14. Diftler, M.A., Mehling, J.S., Abdallah, M.E., Radford, N.A., Bridgwater, L.B., Sanders, A.M., Askew, R.S., Linn, D.M., Yamokoski, J.D., Permenter, F.A., Hargrave, B.K.: Robonaut 2—the first humanoid robot in space. In: IEEE International Conference on Robotics and Automation, pp. 2178–83. Shanghai, China (2011)
15. Araque-Isidro, J.E., Cafolla, D., Ceccarelli, M.: Problems and requirements for outer space astronaut service robot. In: Advances in Italian Mechanism Science IFToMM Italy 2022. Mechanisms and Machine Science, vol. 122. Springer, Cham (2022)
16. Ceccarelli, M., Li, H., Carbone, G., Huang, Q.: Conceptual kinematic design and performance evaluation of a chameleon-like service robot for space stations. Int. J. Adv. Robot. Syst. **12** (2015)
17. Cafolla, D., Ceccarelli, M. Dimensional design of Torveastro, a space orbital station service robot. In: Proceedings of Torveastro Conference 2023. Springer, Cham (2023) (submitted)
18. Robotis.: Dynamixel 64 Manual. https://emanual.robotis.com/docs/en/dxl/mx/mx-64/. Accessed 22 Nov 2022
19. Raspberry pi documentation.: Raspberry Pi hardware. https://www.raspberrypi.com/documentation/computers/raspberry-pi.html. Accessed 22 Nov 2022
20. Robotis: U2D2 Manual. https://emanual.robotis.com/docs/en/parts/interface/u2d2/. Accessed 22 Nov 2022

Check for updates

Materials for Space Astronaut Service Robots

Leandro Iorio(✉), Alice Proietti, Fabrizio Quadrini, and Loredana Santo

Department of Industrial Engineering, University of Rome Tor Vergata, 00133 Rome, Italy
{leandro.iorio,alice.proietti,fabrizio.quadrini,
loredana.santo}@uniroma2.it

Abstract. The importance of materials selection for advanced space applications is mainly due to the harsh space environment; indeed, high thermal gradients, several cosmic radiations, vacuum as well as orbital debris are just some of the factors that must be considered while dealing with in-space applications. An interesting application is the astronaut service robot. International space agencies all over the world are moving toward the extensive use of such robots to perform extravehicular activities (EVA) on orbiting space station, as in the International Space Station (ISS). Traditional structural materials such as Aluminum and Titanium alloys and steels are the most reliable solutions for this kind of applications, while copper-based alloys are more suitable for the manufacturing of electrical and electronic parts. More recently, advanced materials such as polymer-matrix-composites (PMC) have been used to fabricate space components, mainly to decrease the payload of the entire structure: moreover, thin polymeric layers can be used for shielding purposes. In this work, both traditional and innovative materials are presented as possible solutions for new astronaut robots manufacturing.

Keywords: Advanced materials · Space environment · Space astronaut service robot

1 Introduction

The use of astronaut robots to perform extravehicular activities (EVA), outside of orbiting stations such as the International Space Station (ISS), requires a careful material choice; high structural performances, resistance to high temperature gradients and good radiations shielding properties are the main properties that these materials should have.

To date, most of the maintenance operations on the ISS are performed by the astronauts, with all the risks arising from operating in harsh environments. Because of this, the development of astronaut robots is of primary interest for the space agencies, in such a way to limit as much as possible human intervention in critical conditions. Therefore, an appropriate materials selection is essential for a correct design and manufacturing of these robots. Anyway, anomalies and failures can always occur after prolonged exposures in open space due to on orbit servicing (OOS) [1], and significant economic losses should be considered. Major aerospace countries moved towards the promotion of commercial implementation of OOS technologies to extend the lifetimes of spacecrafts [2] and robots. On-orbit servicing surround a huge set of activities performed after spacecraft

© The Author(s), under exclusive license to Springer Nature Switzerland AG 2023
M. Ceccarelli et al. (Eds.): TORVEASTRO 2023, MMS 130, pp. 100–109, 2023.
https://doi.org/10.1007/978-3-031-28447-2_8

deployment, such as maintenance, repair, assembly [3], capture of cooperative and non-cooperative objects, refueling, deployment, extravehicular activity supports, and debris removal [2]. Since these activities are time consuming and dangerous, space robotics systems represent one of the most effective solutions to perform these tasks. In addition, all the facilities required by human presence in open space, such as life-support devices, are not necessary for robots [4].

The mains topic for actual research concerning space robotics technologies for OOS deal with planning and control algorithms [2] as well as the structures and testing technologies which are central prerequisites and guarantees for algorithm implementation. Design and manufacturing technologies play a central role in the development and manufacturing of reliable and suitable in-space facilities and materials selection is a key factor. Environmental factors and manufacturing requirements dictate the guidelines for astronaut robots' manufacturing.

In this chapter, both traditional and innovative materials are discussed as a possible solution for manufacturing new astronaut service robots.

2 The Space Environment

The scientific community has shown a great interest toward the space environment from space living to space observation; the research attention is manly focused on areas such as medicine, material science, technology developments and human sciences. The interests of scientists and researchers all over the world towards these topics become stronger every day. Considering this, a lot of materials have been studied over the years and are currently others are still under investigations to be used in in-space advanced applications. Materials with specific characteristics in terms of lightness, high melting temperature, high corrosion resistance are always central points in these kind of investigations [5, 6]. A deep understanding of the environmental characteristics is fundamental to overcome the strong limitations associated for these applications. Many aspects must not be neglected when looking at the great potential and important insights that move and come from this relatively new environment.

The near-earth space environment is of great interest, and it is an active zone for aerospace industry. In this area, devices are exposed to several sources of damage such as high thermal gradients, vacuum, erosion, impacts from micro meteors or debris, radiations as UV or space plasma. For this reason, aerospace components should be designed to resist to all these factors, that cause the weakening of their mechanical, chemical, and shielding properties. Moreover, electronic, communication and control systems, can be negatively influenced by all those forms of damage [7].

The first problem dealing with materials expected to work for long time in space environment is vacuum. Long time exposure is responsible of outgassing from the surface of the material which in turn lead to contamination of surfaces, loss of dimensional stability, as well as harmful effects on the properties of materials. If contaminant fixes on the surface of the material and interacts with other spatial environmental aspects, such as Ultraviolet (UV) radiation, thermal cycles or atomic oxygen, management of the contamination become a complex task [8].

The main source of danger to consider in space environment regards the presence of radiations, differing in nature and composition [9]. In fact, the structures and the behavior of materials can be seriously affected. Historically, radiation is classified depending on the nature of the source; three types of radiation can be identified: galactic cosmic rays (from outside of the galaxy), solar particle radiation (within the solar system), and trapped particle radiation (within the Van Allen Belts). Galactic cosmic rays (GCR), Solar cosmic rays and Earth's radiation belts are the first radiations to consider. The main constituents for GCR are protons (85%) and α-particles (about 13%). Solar particle events such as coronal mass ejection and flares give rise to solar cosmic rays, which are mostly constituted of protons/electrons and a small percentage of heavy ions. High energy electrons, protons, α-particles, heavy ions, and high energy photons are the main constituents of earth's radiation belts [10]. Operating in harsh space environment and radiation exposure lead to mechanical, chemical and morphological deterioration of the materials. These effects are the direct consequence of the energy transfer during the interactions, that causes ionization, phonon excitations and atomic displacement. The energy transfer is so detrimental because the magnitude of radiation energy varies from kiloelectron volt for the Van Allen Belts to giga-electron-volt for galactic cosmic radiation and solar heavy ions [11]. Organic materials are subjected to all these detrimental effects, thus the improvement of shielding properties is necessary. These forms of radiation are experienced by all space structures operating on low Earth orbit (LEO) at different degrees, depending on both the period of the solar cycle and on the altitude, inclination and duration of the orbit.

Moreover, solar electromagnetic radiations coming from energy output of the sun in the form of electromagnetic waves are also present. More in detail, UV irradiation nearby the sun is an electromagnetic (EM) radiation with a wavelength of about 200–400 nm. UV energy as a function of the orbit and the exposure time may be sufficient to interact with materials not only on a macroscale but also at a microscale. For example, in organic materials it is able to break the chemical bonds. Specifically, Table 1 summarizes most of the common radiation crossing the space environment.

Another aspect to be considered before materials choice for aerospace structures and facilities is thermal excursion. In fact, solar activities cause temperature changes between day and night. Moreover, temperature gradients can develop inside materials as a function of the different orientation of materials surfaces towards cosmic rays leading to thermal fatigue [12]. Also, thermal cycles are influenced by the orbit, in a 90 min LEO cycle, the temperatures detectable on the surfaces of a spacecraft range from −100 °C to −90 °C, with highs of −120 °C to 120 °C.

In geostationary orbit, the excursion is greater between −150 °C and 150 °C. Considering this, thermal stability and non-flammability are fundamental requirements for materials selection.

Moreover, the presence of atomic oxygen [13] must be considered in the LEO, and it greatly affects the superficial optical properties of any material. Complex physical and chemical reactions develop among atomic oxygen and the surface of materials. Consequently, materials damages, such as cracking and loss of quality, changes in surface optical properties, may occur.

Table 1. Main source of radiation in aerospace environment

Identified source of radiations	Composition
Cosmic rays	85% protons
	12.5% α-particles (he-ions)
	1% heavy nuclei
	Small flux of energetic gamma ray photons
Solar cosmic rays	Mostly protons/electrons
	Small percentage of heavy ions
	Flux of X-rays
Van Allen belts	Protons and electrons + alfa particles
Solar electromagnetic radiations	UV irradiation

Orbital debris, that are objects of different sizes and shapes which can originate from collisions and impacts of satellites and body of rocket, are another source of damage. The amount of orbital debris has become a urgent problem to solve; this is due to the fact that the number space missions has recorded an impressive growth in the last decades [14]. Despite the preventive measures that are going to be adopted to mitigate their effect, a linear growth of their amount in the next 100 years can be predicted. Other problems arise from the difficulties in tracking all the objects especially the ones with dimensions lower than 1 cm. For this reason, many efforts have to be made in order to develop new and advanced tracking strategies [15]. Accordingly, materials operating in space are required to exhibit considerable mechanical properties to limit damages which should occur in case of collision with orbital debris.

3 Materials for Aerospace and Space Robotics

Materials which could be used for astronaut robots get the same that are used for aerospace structures being manned or unmanned, in low Earth to geosynchronous orbit, cis-lunar, lunar, planetary or deep space exploration [16].

NASA criteria about flammability and outgassing change depending on the environmental condition that the material have to withstand. In particular, outgassing or molecular contamination can impact spacecraft performance, as in the case of the Stardust spacecraft [17]. To minimize molecular contamination through materials selection and thermal vacuum bakeout the ISS plan, dealing with the material exposure on space environment helps in the choice of the correct material which guarantees an established level of contamination, and which has been set to 130 Å. Moreover, outgassing could be dangerous also inside the spacecraft, as in the case of the Environmental Control and Life Support System (ECLSS) on ISS, where silicone outgassing hampered the performance of beds used for carbon dioxide removal.

Generally, the strength-to-weight ratio drives the choice of structural materials for aerospace structures. The presence of static and dynamic load must be considered as well as thermal corrosion protection, manufacturability, repairability and cost. Moreover, use-dependent properties must be also considered, such as dielectric constant for radomes or gas permeability (for tanks). In addition, several structures can require high tolerances in coefficient of thermal expansion, such as telescope optical benches; in this case, the choice of materials moves toward composite rather than traditional metallic materials. If a high toughness is required, metallic matrix composites are particularly suitable. Atomic oxygen erosion limits the use of non-metallic materials especially in low Earth orbit, where particle radiation and ultraviolet rays can break the cross-linking of polymeric materials. In fact, long term exposure had induced satellite lost of a full ply due to atomic oxygen [18]. Embrittlement and strength loss in some organic materials are the main risks induced by high radiation environment.

Stiffness is another fundamental property for space devices. Honeycomb structures greatly fulfill this function and can be with wither composite or metallic face sheets and cores. Passive thermal control is another aspect to consider for materials to be use in aerospace; it can be reached by surface treatment, application of a coating or covering with a multi-layer insulation blanket. For example, before launch, surface treatment is performed on metals to avoid corrosion. Thermal control is even more stringent for those structures intended for EVA, since thermal excursion can vary in the range -118 °C + 113 °C. If not, it can be pursued by using coatings or paints, and four kinds of binder are qualified for space applications being polyurethane, epoxy, silicone or potassium silicate. Among these, epoxy and polyurethane-base coatings have limited life in LEO since the action of atomic oxygen erodes the binder leaving only pigment particles behind [19]. On the other side, potassium silicate is difficult to apply and is sensitive to contamination, but it has the advantage of proven durability in space and it is considered safest around optics [20]. Multilayer insulation structures can be manufactured by thin polymeric films.

Thermal protection materials differ from thermal control materials since the former are used for high temperature applications characteristics around the engine, while the latter are used to moderate on-orbit temperature effects. Reusable materials such as tiles or ceramic-matrix composites can be used for heatshields. On the Space Shuttle silica ceramic tiles were used. They were lightweight and can be formed in various densities; moreover, they can handle reentry temperature up to 1260 °C, but in turn are fragile and easily damaged.

According to the discussed different forms of space radiations, shielding is another important property to guarantee. Low Z materials (being materials with low atomic number) such as polyethylene, protect from beta particles (mainly electrons). Composite with a low and high Z materials, can improve shielding per areal weight performance for shielding electronics. In this case, optimal radiation shielding for long-term manned mission still needs to be developed. Hydrogenated graphite nanofiber and other nanocomposites are the most promising solution.

To limit damages due to impact with orbital debris, the Whipple shield is the actual adopted solution. In detail, the sacrificial bumper, breaks up the impacting particle into smaller pieces or, in case of extremely high impact velocities, vapor. Different materials can be used as bumper, depending on the kind of mission; in the case of Deep impact mission, the impactor has five sheets of copper. The need to ensure both thermal protection and hypervelocity impact protection lead to the production of a multi-layer insulation blanket.

Finally, optical properties should be maintained with the use of appropriate materials in that space radiations can react with the impurities content of material used for windows. For this reason, coating the window with cerium oxide will also improve the durability in space.

Materials for Aerospace applications can be broadly divided into four classes: metallic materials (metallics), non-metallic or polymeric materials, composite materials (composites) and ceramic materials (ceramic).

3.1 Traditional Structural Materials

Traditional structural materials used for Aerospace applications are metals like aluminum alloys, titanium alloys, iron alloys (steels) and copper alloys [16].

Aluminum alloys are generally the most used material for structural parts in aerospace. In detail, wrought alloys, produced by casting plus deformation, are preferred for most load bearing structures because of their high strength and damage tolerance. On the other hand, aluminum alloys produced by casting can be used to produce intricate shaped parts; in this case, they have limited strength and ductility. Several aluminum alloys have been developed and the classification schemes can be easily found in the ASM handbook.

The most commonly wrought aluminum alloys used for aeronautical and aerospace applications are 2xxx and 7xxx series, where the difference is related to the amount of the first alloying element being copper for the former and zinc for the latter. Aluminum alloy of the 2xxx series have lower crack growth rates and better fatigue resistance than 7xxx series alloys, and are used on lower wing and body skin. In detail, Al 2014A is a high strength alloy which maintains this property also after heat treatment. It has a good machinability that make it particularly suitable for the Defense field. Typical applications are the production of high-strength components, aircraft structures and fittings.

The 7xxx series alloys offer the greatest potential in terms of mechanical properties. The oldest one used is the 7075-T6, the other ones introduced showed higher mechanical properties, for example the 7178-T6 has higher strength and is used in structures requiring compressive loading.

Moreover, 7075-T74 was developed for thick section applications and mostly used in aircraft applications; it shows high strength, good stress corrosion cracking resistance and good fracture toughness and fatigue resistance. 7150-T6 alloy plates and extrusion were used as wing skins and stringers in the upper wing structures. Notable advances have taken place in the development of newer 7xxx series alloys including 7090, 7091 and CW67 where the strength, toughness and stress corrosion cracking resistance are improved. In detail the CW67 offers the best combination of strength and fracture toughness indicating significant potential for weight savings. Also cast aluminum alloys are available as 200series, 300series, 400series and 500series. Historically aluminum alloy castings have been used in nonstructural parts in aircraft including pulley brackets, quadrants, doublers, clips, ducts and wave guides. They also have been used extensively in valve bodies of hydraulic control systems.

Titanium is the fourth most abundant structural metal and the high strength, low density, excellent corrosion resistance, and biocompatibility are the main characteristics that make it attractive for a variety of applications (aircrafts, aero-engines, biomedical devices and chemical processing equipment) [21]. The major use is in airborne applications such as aero-engines. Airframes, missiles and spacecraft. It has the highest strength-to-density ratio, but it is used only for niche applications, because of the high cost due to the reactivity of titanium with oxygen and high raw materials. Titanium can be used at temperatures higher than that allowable for Aluminum but its high reactivity with the oxygen reduces the operating temperature at about 600 °C. The oxide layer developing after reaction with oxygen allow the enhancement of corrosion resistance in aggressive environments. Alpha + beta alloys are the one used in the greatest quantity and are applied when high strength fracture resistance, fatigue strength, elevated temperature capability with good corrosion resistance are needed (aircraft structures and spacecraft components). The most used titanium alloy is Ti6Al4V which combines the good mechanical strength and low density with the high corrosion resistance. It can be thermically threated also in the case of thick section (up to 25 mm). It is generally used in the manufacturing of engines.

Steels are the most common metals used in structural applications because they have high strength and ductility and relatively low cost. One drawback is density, being significatively higher compared to aluminum and titanium alloys. The good corrosion resistance makes the stainless steel, as the 310, 302, particularly suitable for these applications. Moreover, Nickel chromium steels are resistant to oxidation and so they are used in propulsion systems. Ultra-high strength steels, as the 4340 are generally used for landing gear and other critical structural members for aircrafts. High Fracture Toughness Steels as the HP 9-4-30 are used for aircraft structural components.

The unique electrical, thermal and wear characteristics make the copper-based alloys particularly attractive for aerospace applications. Apart of using copper alloys such as Cu-Cr or Cu-Zr in regeneratively cooled rocket engines, copper alloys can also be used in electronics and electrical systems. Cu-based alloys are commonly used for cold plates and thermal back planes of standard electronic modules whereas electrical wiring, electrical motors, electrical actuators generally use Cu wire.

3.2 Polymers and Composites Materials

Polymers are organic compounds being chemically based on carbon, hydrogen and other non-metallic elements (O, N, Si) [16]. They have low density, thermal conductivities, and moduli and can be used in light-weight applications. The addition of thermally and/or electrically conductive fillers make polymers suitable in application where isolative or conductive requirements are needed. For this reason, lightning strike, electromagnetic interference (EMI) shielding, and antistatic protection applications are common fields of use of polymers. Polymeric matrix composites can resist to several mechanical cycles during aircraft lifecycle. Many aerospace structures must also be tolerant of high temperatures, moisture levels, thermal cycling and UV-radiation. In aerospace applications, thermosetting polymers are used where higher temperatures are expected. Moreover, mechanical properties can be greatly enhanced when using thermosetting based systems reinforced with silicon carbide whiskers [22]. The most common used thermosets are epoxies, bismaleimides (BMI), cyanate esters (CE), phenolics and polyimides (PI). The use of epoxies and BMI is limited by their susceptibility to brittleness and moisture attack. On the other side, CE resins are more easily processed and offer excellent strength, moisture absorption and electrical properties. PI resins have good adhesion and heat and chemical resistance as well as superior mechanical properties. Thermoplastic polymers are a good alternative to thermosets if high temperature resistance is not required. Thermoplastics have a much higher impact strength than thermosets. Commonly used thermoplastics in aerospace are polyetheretherketone (PEEK) which has excellent stiffness, good resistance to moisture attack, corrosion abrasion and chemical and radiation exposure. Other common used resins are polyetherimide (PEI), that has high heat resistance, strength and modulus, poly(p-phenylene sulfide) (PS) and thermoplastic polyimides which offer excellent performance at elevated temperature. The most used fiber reinforcements are carbon and aramid. In Table 2 the commonly used materials for aerospace applications are summarized.

Among composites, another class of materials to consider are smart materials. They are able to react at stimuli coming from the external environment, and can be used as actuators in the development of soft robotics [23]. Shape memory polymer composite (SMPC) materials belong to this class; they can be used as actuators thanks to the possibility of being activated through simple electrical devices and can also exert high forces [24]. SMPC have been tested in the space environment, showing their capability of using in space [25, 26].

4 Material Proposal for Astronaut Robot

Literature review analysis suggests that the most reliable solutions for aerospace structures, including astronaut robot for EVA, are still the traditional structural materials. Al alloys and Ti alloys offer the optimal compromise between mechanical properties and low weight requirements.

Table 2. Materials commonly used for aerospace applications

Class of materials	Composition
Metallics	Al alloys
	Ti alloys
	Steels
	Cu alloys
Composites	Thermoplastic based composites (PEEK, PPS, PEI, PI)
	Thermosetting based composites (epoxies, BMI, CE, PI)

The actual robot used for EVA on ISS as the robotic arm produced by Canada consists of stainless steel, titanium alloys and carbon fiber reinforced composites with epoxy matrix [27]. The same choices have been done for Canadarn 2 being the evolution of the former robotic arm, with the only difference of using a thermoplastic-based matrix instead of thermosetting-based one. The European robotic arm (ERA) [28], launched in 2021 and used for the logistic in the Russian segments is also made with fittings in titanium and rods in carbon fiber reinforced composites. The harsh environment typical of the LEO requires proper thermal insulation to guarantee the proper functioning of the electronics, and this was achieved in the Canadarn thanks to a multilayer structure of Kapton, fabric Draco and a glass fiber fabric.

Ti and Al alloys described in the previous section are also used as structural materials for the new Rover Perseverance by NASA that would operate in environment even more severe than LEO being the Martian surface [29]. This confirms that in advanced applications as the one regarding the aerospace ones, the traditional materials are the best choice especially for the reliability they offer. New materials are anyway under examination thanks to the Materials on International Space Station Experiments (MISSE) experimentation; its aim is to evaluate the behavior of different class of novel materials in open space for a long time.

5 Conclusion

In conclusion, till today the best choice for manufacturing astronaut service robots is the use of traditional Al e Ti alloys as regarding fittings, end effectors and frame. The connecting rods between the end effectors and the central body can be made in thermosetting-based composites with carbon fiber reinforcements. SMPCs could be used as actuators, as above mentioned. Moreover, polymeric-based materials in the form of thin film can be used for thermal insulations and shielding from different cosmic radiation. Further studies must be performed to deepen this topic mainly taking into account the problems and requirements for outer space astronaut service robot [30].

Acknowledgements. This work has been supported by grant Lazioinnova-POR214–2020-A0375–2020-36517 for the project TORVEASTRO that is gratefully acknowledged.

References

1. Akin, D., Sullivan, B.: A survey of serviceable spacecraft failures. In: AIAA Space 2001 Conference and Exposition. Albuquerque 4540 (2001)
2. Li, W.J., Cheng, D.Y., Liu, X.G., et al.: On-orbit service (OOS) of spacecraft: a review of engineering developments. Prog. Aerospace Sci **108**, 32–120 (2019)
3. Liu, H.: An overview of the space robotics progress in China. In: Proceedings of the International Symposium on Artificial Intelligence Robotics and Automation in Space, pp. 15–20. Montreal (2014)
4. Bekey, G.A., Ambrose, R., Kumar, V., et al.: Robotics: State of the Art and Future Challenges. Imperial College Press, London (2008)
5. Holt, D.E.H., Holt, D.: Materials technology for aerospace applications. Mater. Des. **6**(1), 18–24 (1985)
6. Kandasamy, J., Kar, V.R., Sultan, M.T.H., Murugan, R.: Materials selection for aerospace components. In: Sustainable Composites for Aerospace Applications (Chapter 1) pp. 1–18 (2018)

7. Louthan M.R.: Materials degradation in space environment. AIAA Special Rep. No. 5 (1980)
8. Chen, J., Ding, N., Li, Z., Wang, W.: Organic polymer materials in the space environment. Prog. Aerosp. Sci. **83**, 37–56 (2016)
9. Chancellor, J.C., Blue, R.S., Cengel, K.A., Aunon-Chancellor, S.M., Rubins, K.H., Katzgraber, H.G., Kennedy, A.R.: Limitations in predicting the space radiation health risk for exploration astronauts. npj Microgravity 8 (2018)
10. Holmes-Siedle, A., Adams, L.: Handbook of Radiation Effects. Oxford University Press (1993)
11. Sajid, M., Chechenin, N.G., Torres, F.S., Khan, E.U., Agha, S.: Space radiation environment prediction for VLSI microelectronics devices onboard a LEO satellite using OMERE-TRAD software. Adv. Space Res. **56**(2), 314–324 (2015)
12. Han, J.H., Kim, C.G.: Low earth orbit space environment simulation and its effects on graphite/epoxy composites. Compos. Struct. **72**, 218–226 (2006)
13. Reddy, M.R.: Effect of low earth orbit atomic oxygen on spacecraft materials. J. Mater. Sci. **30**, 281–307 (1995)
14. Gover, J.E.: In: IEEE 1980 Annual Conference on Nuclear and Space Radiation Effects, Cornell University, Ithaca, NY, 15–18 July 1980
15. Cament, L., Adams, M., Barrios, P.: Space debris tracking with the poisson labeled multi-bernoulli filter. Sensors **21**, 3684 (2021)
16. Bhat, B.N.: Aerospace materials and applications (2018). https://doi.org/10.2514/4.104893
17. http://stardust.jpl.nasa.gov/news/vision.html
18. Ann, F., Young, W., Young, L.: An overview of the first results on the solar array materials passive LDEF Experiment (Sample), AO171. In: LDEF First Post-Retrieval Symposium, NASA CP-3134, June 1991
19. Stein, B.: LDEF materials: an overview of the interim findings. In: LDEF Materials Workshop, NASA CP-3162, Nov 1991
20. Finckenor, M., Pippin, H.G., Frey, G.: MISSE thermal control materials with comparison to previous flight experiments. In: 9th International Space Conference—Protection of Materials and Structures from the LEO Space Environment, Toronto, Canada, 20–23 May 2008
21. Properties and selection: nonferous alloys and special purpose materials, vol. 2. ASM Handbook (1992)
22. Avella, M., Dell'erba, R., Martuscelli, E., Partch, R.: Thermosetting based composites reinforced with silicon carbide whiskers. J. Polym. Mater. **17**(4), 445–458 (2000)
23. Hao, Y., Zhang, S., Fang, B., Sun, F., Liu, H., Li, H.: A Review of smart materials for the boost of soft actuators, soft sensors, and robotics applications. Chin. J. Mech. Eng. **35**(1), 1–16 (2022). https://doi.org/10.1186/s10033-022-00707-2
24. Quadrini, F., Iorio, L., Bellisario, D., Santo, L.: Shape memory polymer composite unit with embedded heater. Smart Mater. Struct. **30** (2021)
25. Santo, L., Quadrini, F., Ganga, P.L., Zolesi, V.: Mission BION-M1: results of Ribes/Foam2 experiment on shape memory polymer foams and composites. Aerosp. Sci. Technol. **40**, 109–114 (2015)
26. Santo, L., Quadrini, F., Bellisario, D., De Groh, K.K.: Shape memory polymer composites and cosmic ray shielding materials in open space. In: Proceedings of the Applied Space Environments Conference (ASEC), Los Angeles, 13–17 May 2019
27. Doetsch, K.H., et al.: https://www.thecanadianencyclopedia.ca/en/article/canadarm
28. https://www.esa.int/Science_Exploration/Human_and_Robotic_Exploration/International_Space_Station/European_Robotic_Arm
29. https://mars.nasa.gov/mars2020/spacecraft/rover/
30. Araque-Isidro, J.E., Cafolla, D., Ceccarelli, M.: Problems and requirements for outer space astronaut service robot mechanisms and machine science. In: MMS, vol. 122, pp. 603–611 (2022)

Check for updates

Additive Manufacturing of Light Alloys for Aerospace: An Overview

Roberto Montanari, Alessandra Palombi(✉), Maria Richetta, and Alessandra Varone

Department of Industrial Engineering, University of Rome "Tor Vergata", 00133 Rome, Italy
{roberto.montanari,alessandra.palombi,richetta,
alessandra.varone}@uniroma2.it

Abstract. The work provides an overview of recent advances in the use of additive manufacturing (AM) technologies for aerospace applications of light alloys by describing current problems and perspectives. The needs of the aerospace industry stem from the strong competition to manufacture aircrafts with improved technical features and reduced costs. In this challenge AM has drawn increasing attention. A critical analysis has been carried out by focusing the attention on the strict relation between process parameters, microstructure and properties of light alloys printed by various AM methods. The matter is discussed by taking into account the aspects common to all the light alloys and those specific of each class of materials.

Keywords: Aerospace · Additive Manufacturing · Light alloys · Aluminium alloys · Titanium alloys · Magnesium alloys

1 Introduction

The term Additive Manufacturing (AM) refers to a set of manufacturing techniques that share one basic principle: the part is produced by adding material in a layer-by-layer deposition. Today, many different AM techniques to print different materials, from polymers to metals, are employed in various fields such as automotive, biomedical, aeronautics and aerospace [1]. These applications have in common the need of tailored high standard components produced in small volumes and in time as short as possible.

The AM technologies used for metals are based on the local melting of a feedstock. As displayed by Fig. 1, the differences among them consist in the feedstock material (powders or wires) and heat source employed to melt the material (laser, electron beam or electric arc). Powders can be either uniformly distributed on the printing plate and then melted according to a CAD file (Powder Bed Fusion-PBF) or deposited through a nozzle where needed and simultaneously melted by a laser source (Direct Energy Deposition-DED). If the feedstock material is in form of wire the material is melted by an electric arc and deposited locally (Wire and Arc Additive Manufacturing-WAAM).

These methods allow to produce metallic components with mechanical properties comparable or better than those of conventionally manufactured ones, and complex optimized geometries. Moreover, AM allows the production of near net-shape-parts with reduction of manufacturing time, material waste and post processing treatments.

© The Author(s), under exclusive license to Springer Nature Switzerland AG 2023
M. Ceccarelli et al. (Eds.): TORVEASTRO 2023, MMS 130, pp. 110–128, 2023.
https://doi.org/10.1007/978-3-031-28447-2_9

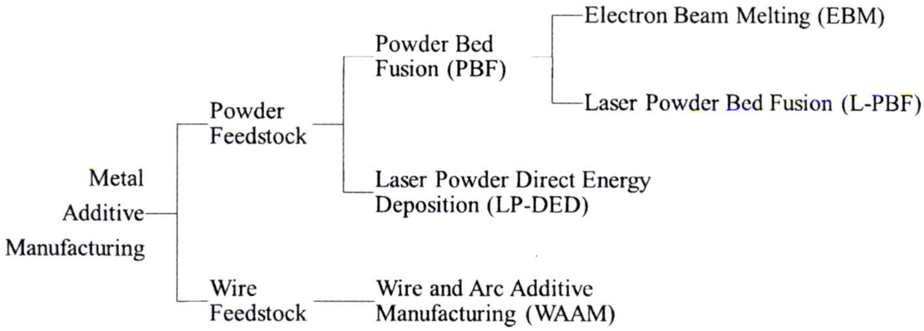

Fig. 1. Schematic view of the main Additive Manufacturing techniques.

Therefore, AM has found increasing application in aerospace industry and light alloys (Al, Ti, and Mg alloys) play a relevant role thanks to their good mechanical properties and low density [2–4].

In aeronautics, AM involves specific benefits mainly connected to the topological optimization (TO) for weight reduction and to the operations of maintenance, repair and overhaul (MRO) including inspection, replacement of damaged parts, refilling of gases and lubricants, coating fixing, etc.

To reduce the weight of aircrafts TO criteria are increasingly implemented in the design and fabrication processes of structural components and this requires the development of greater computational power, more sophisticated softwares, the integration of robotics in machining processes, and novel manufacturing technologies such as AM. Therefore, the main companies of aircraft industry have been exploring the possibility to produce through AM components such as interior parts, brackets, hinges, and also fuselage and airframe of lighter weight. Some examples of application of TO methods for reducing the mass of aerospace components are shown in Fig. 2.

RUAG successfully manufactured an antenna bracket (b) that has been certified for use in outer space (0.94 kg instead of 1.6 kg of the original part) [5, 6]. The weight reduction of radio frequency filter printed by Airbus Defense and Space, and used in satellites (d), was 50% [7]. Thales Alenia Space manufactured a Ti antenna bracket for a geostationary telecommunication satellite (c) 25% lighter than the original one with reduced production time and costs [8].

In addition to structural parts AM applications are also related to the engines: integrated piping systems, compressors, fuel nozzles and turbine blades with internal cooling channels [10–12].

An aspect of interest related to MRO operations is the just in time AM production of mechanical parts based on digital data through relatively simple machinery. This could be very useful for the supply chain even if certification standards still represent a serious obstacle to advancements in this direction.

Light alloys prepared by AM exhibit some common microstructural characteristics (columnar grains and second phases out of thermodynamic equilibrium), similar defects such as porosity and surface of poor quality, anisotropic properties with improved strength respect to corresponding materials prepared by casting but reduced toughness,

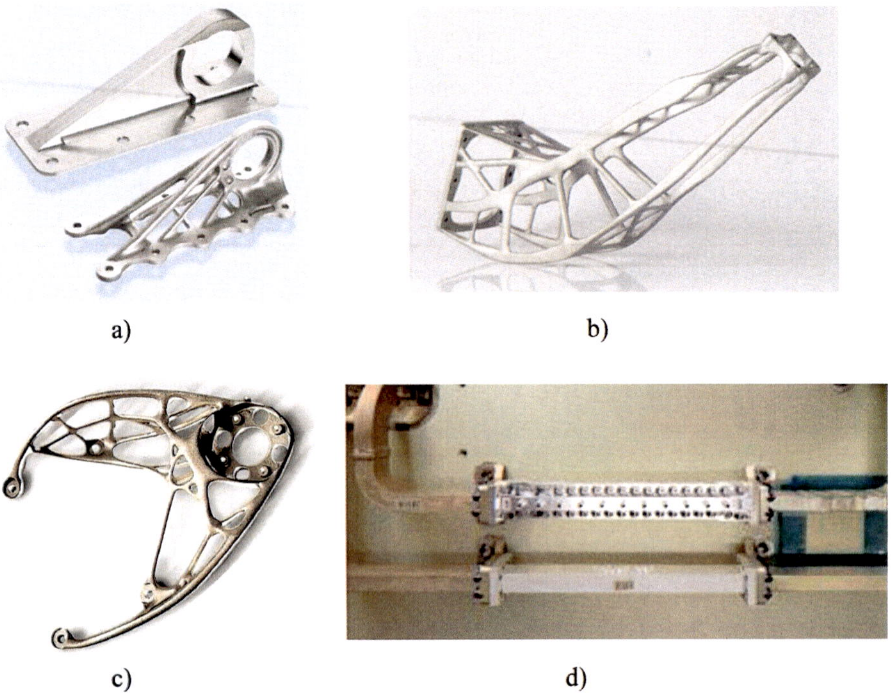

Fig. 2. Ti6Al4V wing bracket of an Airbus A350 XWB [9] (a); AlSi10Mg antenna bracket for RUAG's Sentinel Satellite [5, 6] (b); Ti antenna bracket manufactured by Thales Alenia Space and 3D Systems [8] (c); AlSi10Mg radio frequency filter manufactured by Airbus Defense and Space [7] (d).

and some residual stresses. On turn, other aspects are specific of each family of alloys and the specific composition.

Aim of this work is to give an overview of AM production of aerospace components focusing the attention on the strict relation between process parameters, microstructure, and properties.

2 Defects and Process Parameters

The most common defects observed in printed components of Al, Ti and Mg alloys are: (i) porosity, (ii) poor surface quality (iii) anisotropy, and (iv) residual stresses.

2.1 Porosity

The main causes of porosity can be found in process parameters, material composition and powder characteristics. The pores can be of different type, namely lack of fusion pores, keyhole pores, gas pores [13, 14]. Lack of fusion pores form when the energy density is too low to completely melt the metal [15], or if the wettability is not optimal [13]. These pores have irregular shapes and, usually, it is possible to identify the un-melted powder particles. On the contrary, keyhole pores occur when the energy density is too high causing turbulences in the melt pool that lead to the formation of pores under the top surface of the melt pool [10]. Gas pores, with spherical morphology, result from gases trapped in the feedstock material (powders or wires) or moisture on the surface of the powders or wires [11]. They are due to a non-optimal choice of shielding gas flux parameters or can originate from the evaporation of low melting alloying elements.

As mentioned before, the choice of process and process parameters can have a huge impact on porosity, surface finish, microstructure, and mechanical properties.

Process-related porosity can be reduced acting on energy density, a factor that depends on scan speed, beam power and hatch spacing. For each material there is a critical value of energy density that minimizes the overall porosity [13, 16–22]. In general, the defects are minimized when a high energy density is employed [23], but in terms of microstructure the result is the increase of grain size and degradation of mechanical properties [24–26]. If the beam power is too high the occurrence of keyhole pores and higher evaporation rate can hinder all the good influence those parameters may have on defect reduction [21, 22, 27].

2.2 Poor Surface Quality

The surface quality mostly depends on process parameters: scan speed, beam power, scan strategy and layer thickness. On the top and bottom surfaces of as-built AM components it is possible to distinguish beam scan tracks and un-melted particles, while melt pools can be observed in the cross-sections. Figure 3 displays un-melted particles on the surface of a Ti6Al4V component printed by L-PBF. Through the control of process parameters melt pool geometry and solidification rate can be suitably changed to obtain components with improved surface quality [28].

2.3 Anisotropy

The thermal history of the material leads to the growth of columnar grains oriented along the printing direction and to an inhomogeneous distribution of porosities and other defects like lack of fusion pores [29–31]. The combined action of columnar grains and inhomogeneous defect distribution is the main cause for anisotropy in mechanical properties of AM components: generally, samples tested along the growth direction present higher Young modulus, yield strength and ultimate tensile strength but lower ductility.

2.4 Residual Stresses

In techniques like L-PBF, high thermal gradients between melt pool and surrounding solid metal result in a rapid solidification that often leads to residual stress, and in some cases, to part deformation [32, 33]. In EBM this phenomenon is less critical due to the pre-heating of the chamber. This solution has been employed also in L-PBF with promising results [34, 35], but the effects on the microstructure are not negligible: lower solidification speed results in coarser microstructure.

Some problems, specific of each class of materials, will be briefly examined in the following.

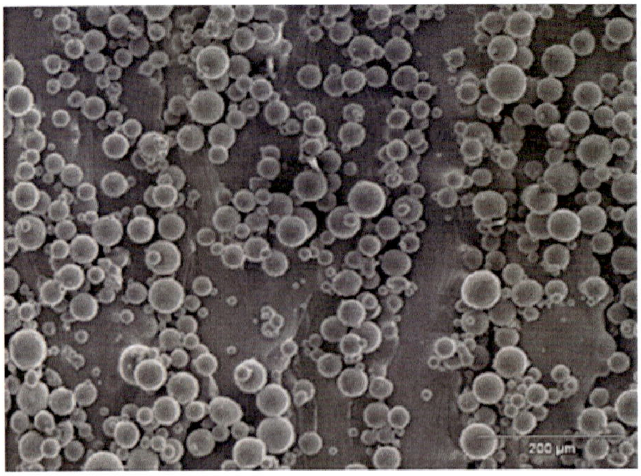

Fig. 3. SEM image of the top surface of a Ti6Al4V sample produced by L-PBF. A lot of un-melted particles cover the surface.

2.5 Al and Al Alloys

Al and its alloys have a high thermal conductivity and high reflectivity in the range of the laser wavelengths used in AM processes: process parameter choice must be a tradeoff between high energy, needed to overcome low absorptivity of Al, and low enough energy to limit keyhole pores and balling defects [36].

To overcome this issue and obtain parts with also better surface quality and geometrical accuracy, Ullsperger et al. [37] proposed a production method that employs ultra-short laser pulses. The final parts presented reduced keyhole porosity, improved relative density, and refined eutectic structure leading to a higher microhardness.

Process parameters that must be tailored, other than energy density, are the those involving the environment in which the printing is carried out, i.e. shielding gas (N2, Ar, air) and gas flux parameters (direction, speed). While powder bed fusion processers are carried out in closed environments, where inert gases fill the printing chamber, in DED or WAAM, a shielding gas flux must be employed to limit oxidation and other possible external contaminations. Tests on samples of 5356 Al alloys, made by WAAM with N2 and Ar as shielding gases, showed how gas choice impacts on part properties [38]. N2 samples present a higher surface roughness and finer grains than Ar samples., however N2 promoted the precipitation of a large number of nitrides with consequent increase of hardness and decrease of ductility.

Another characteristic to be considered is oxide formation on both the feedstock material and the surface of melt pool during the printing process. The high beam density can be responsible for the disruption and partial evaporation of the oxide layer, the oxide particles that remain in the melt pool may influence the wettability and the amount of trapped gas leading to a higher degree of porosity [36].

Some studies on Al alloys showed how the anisotropy of mechanical properties [16, 39–41] is caused by microstructural inhomogeneity both in terms of different grain morphology along building and normal directions (columnar grains) and of defect distribution [30, 31].

Corrosion resistance is also anisotropic; for example, Al12Si printed by L-PBF exhibits in a solution of NaCl a higher corrosion resistance on the surface oriented along the printing direction [42].

Another relevant issue in the AM of Al alloys is hot cracking caused by shrinkage during the solidification of liquid metal trapped between already solidified dendrites. Crack mainly form parallel to the building direction [43, 44] with the possibility of successive propagation along grain boundaries [45].

One of the most effective methods to limit hot cracking is the modification of the alloy composition to get the solidification in a larger temperature range and the formation of precipitates. For example, Sc and Zr give rise to the formation of Al_3(Sc, Zr) precipitates that promote an equiaxed structure [46–51]; other elements like Si give rise to the eutectic structure [52].

2.6 Ti and Ti Alloys

In Ti alloys the microstructure variations in different directions are more evident than in Al alloys involving a great anisotropy of mechanical properties: yield strain and UTS along the building direction can be 20–30 MPa higher than in the perpendicular direction [53].

AM techniques, for Ti alloys, can lead to beneficial effects in reduction of typical defects like the segregation of β-flecks, generally found in conventionally manufactured component. This kind of defects, caused by the segregation of β-stabilizing elements, is limited by the peculiar thermal cycle the material undergoes in AM processes (especially EBM) favouring diffusion and homogenization of the alloy composition [54].

2.7 Mg and Mg Alloys

Typical AM defects can be found in Mg alloys as well. Moreover, in age hardenable Mg alloys, like WE43 (Y 3.70–4.30, Zr 0–0.4, rare earths 2.4–4.4 wt.%), the thermal cycle of AM processes induces the formation of a fine dispersion of β1(Mg3Nd) precipitates. While the typical ageing sequence is β" → β' → β1 → β, in L-PBF the formation of β1 occurs by-passing the formation of β" and β'. Solution annealing plus aging treatments lead to the growth of a plate-like structure from this phase, instead of its transformation into the equilibrium β (Mg14Nd2Y) phase. This latter transformation, according to Hyer et al. [14], is inhibited because of the formation of a fine dispersion of stable Y2O3 phase in the Mg matrix.

3 Microstructure and Mechanical Properties

The thermal history typical of metals processed by AM generally leads to the formation of columnar grains, oriented along the printing direction, with a fine microstructure and the presence of metastable phases [13, 55–59]. This microstructure is responsible for the good mechanical properties of printed materials: Young modulus and yield strength are commonly higher than those of the same alloys conventionally manufactured [35, 60–63], even if Young modulus is strongly affected by the degree of porosity [64, 65] and porosity and poor surface roughness cause lower fatigue resistance [66, 67]. Several studies have been carried out to find strategies promoting the columnar to equiaxed grain transition, e.g. by adding inoculants [50, 68, 69] or by tailoring process parameters [70].

3.1 Al and Al Alloys

The typical microstructure of Al alloys consists of Al grains elongated along the building direction with a fine distribution of precipitates decorating the grain boundaries and the cells inside the grains [55, 56, 71–73]. An example of such features is shown in Fig. 4 for the AlSi10Mg alloy. The morphology, size and orientation of grains and cells is generally homogeneous from top to bottom of printed parts [74].

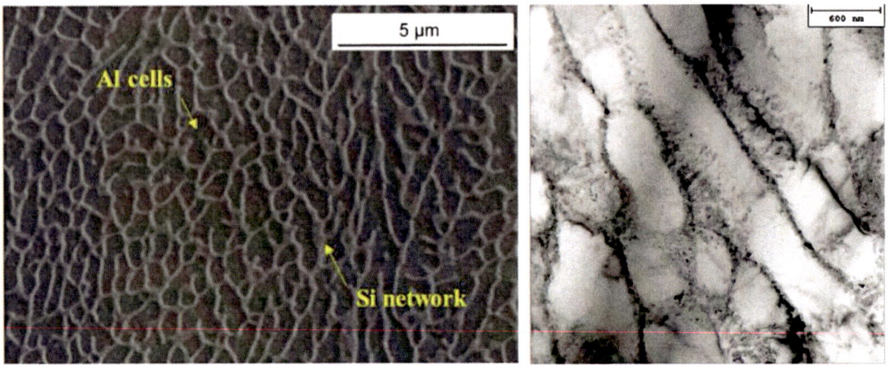

a) b)

Fig. 4. AlSi10Mg alloy printed by L-PBF. Network of cells surrounded by small Si precipitates (a) which are better highlighted by the TEM micrograph at higher magnification in (b).

Repeated laser or electron beam passes act as a cyclic heat treatment that induces the precipitation of different compounds depending on the alloy composition.

Equiaxed grains can be obtained with the addition of specific inoculants [68, 69]. There is a narrow window of process parameters to obtain equiaxed grains and the building direction has a high impact on this aspect. For instance, Hadadzadeh et al. [75] found that the amount of columnar grains is greater in samples with the large dimension oriented along the printing direction.

Many studies have been focused on the optimization of precipitate distribution to obtain fracture toughness values meeting the high aerospace standards. Among the AM techniques, WAAM allows to print alloys with complex composition, e.g. Al-Zn-Mg-Cu, and achieve high specific strength, good fracture toughness and excellent fatigue resistance [76–79].

Another relevant aspect involved in AM printing of Al alloys is the role played by inclusions. Al has a high reactivity with O [80] thus if the shielding gas (WAAM) or the chamber environment (L-PBF and EBM) are not good enough to prevent this reaction, oxides may form and remain trapped in the solidifying metal forming inclusions [81] with detriment to ductility and toughness.

3.2 Ti and Ti Alloys

Large part of studies on AM printed Ti alloys have been made on Ti6Al4V, a $\alpha + \beta$ alloy. The microstructure forms from prior columnar β grains, oriented along the printing direction, and, depending on the solidification rate, may consist of fully martensitic α' (e.g. see Fig. 5), lamellar $\alpha' + \alpha + \beta$ or ultrafine $\alpha + \beta$ [13, 57, 82, 83].

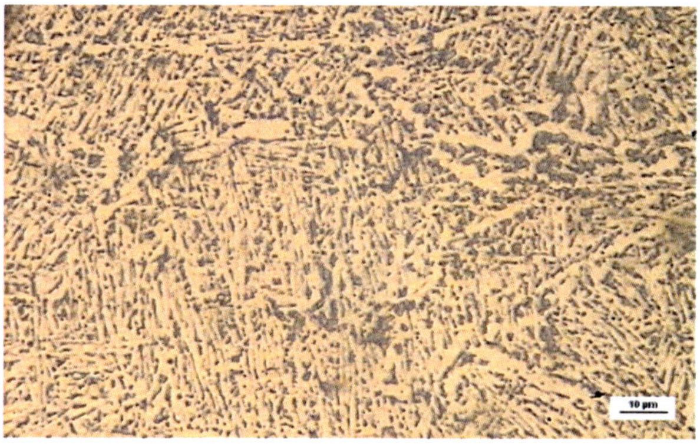

Fig. 5. Martensitic structure of Ti6Al4V printed by L-PBF.

The printing process affects the solidification rate and consequently the parts can present a graded microstructure. In EBM there is a variation of the thermal gradient from the building platform to the top of the built thus near the building platform the grain are coarser than those of the last printed layers [84]. This phenomenon has been observed also in samples produced by DED with an inverse trend: grains near the building platform are smaller than those on top of the part [85]. The reason beyond this phenomenon is that in EBM the printing chamber is held at high temperature, powder is spread on layers that completely cover the printing area and the building plate is pre-heated, while DED is usually carried on in air, the powder is sprayed directly where needed and the building platform is not isolated or pre-heated.

The formation of this graded microstructure can be prevented controlling the melt pool size: studies on WAAM [86], L-PBF [25], and DED [26] showed that constant energy density results in homogeneous grain size along the component.

In Ti alloys microstructural anisotropy can be found in samples with the main dimension directed along the building direction or perpendicularly: in both cases there is a columnar structure, but in the first case the microstructure is finer than in the latter one [87].

3.3 Mg and Mg Alloys

Microstructural features of Mg alloys produced by AM are quite different from the ones discussed for Al and Ti alloys. In this case the as built microstructure presents a mix of columnar and equiaxed grains which tend to coarsen when beam energy increases [88, 89]. As shown in Fig. 6, the WE43 alloy produced by L-PBF exhibits equiaxed grains in the center of the melt pool that gradually transform into a lamellar structure [69].

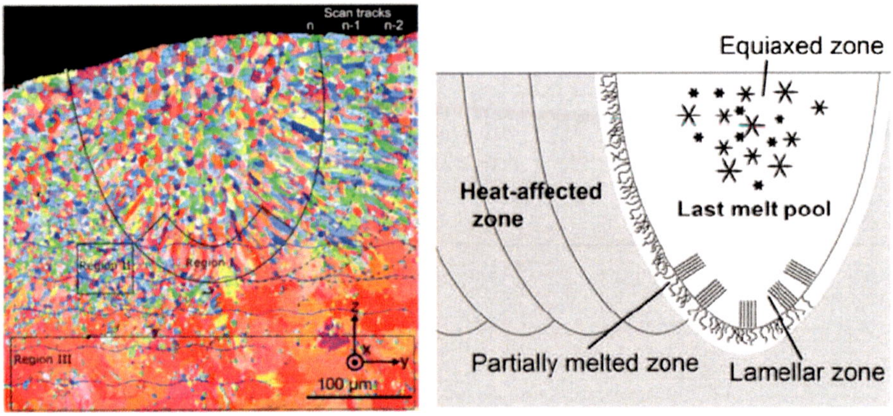

a) b)

Fig. 6. (a) EBSD map of the last laser-scan track of WE43 Mg alloy produced by Laser-PBF; (b) scheme of the different types of microstructure: (i) partially melted zone (ii) lamellar zone and (iii) equiaxed zone. (from Ref. [58]).

One of the main issues in the printing of Mg alloys is the high Mg evaporation rate that can significantly change the alloy composition and can give rise to relevant gas porosity [66, 89]. Improvements can be achieved by increasing the pressure in the building chamber shifting the boiling point to higher temperatures and limiting evaporation [90].

Parts made through WAAM exhibit a graded microstructure: near the printing base, where the thermal gradient is higher, the microstructure presents columnar grains; in the following layers there is a transition to equiaxed microstructure and in the top layers equiaxed and dendritic grains are observed. This inhomogeneity leads to an anisotropy in mechanical properties: specimens tested horizontally presented a higher yield and ultimate tensile strength [91].

4 Post Process

Given the relatively high porosity and poor surface quality, AM components generally present low fatigue resistance. Hot Isostatic Pressing (HIP) and surface finishing have been proved to be valid solutions for improving fatigue behavior [14, 82, 92–94].

Moreover, conventional heat treatments can be performed on printed materials to induce precipitation and reduce residual stresses [35, 95, 96].

4.1 Al and Al Alloys

Heat treatments can influence density and size distribution of the pores [97]: if the materials are not preventively submitted to HIP, unsolvable gases trapped in the pores expand during heating. To prevent pore expansion, it is a good practice to perform HIP before heat treatment, or even both of them in one single step [98].

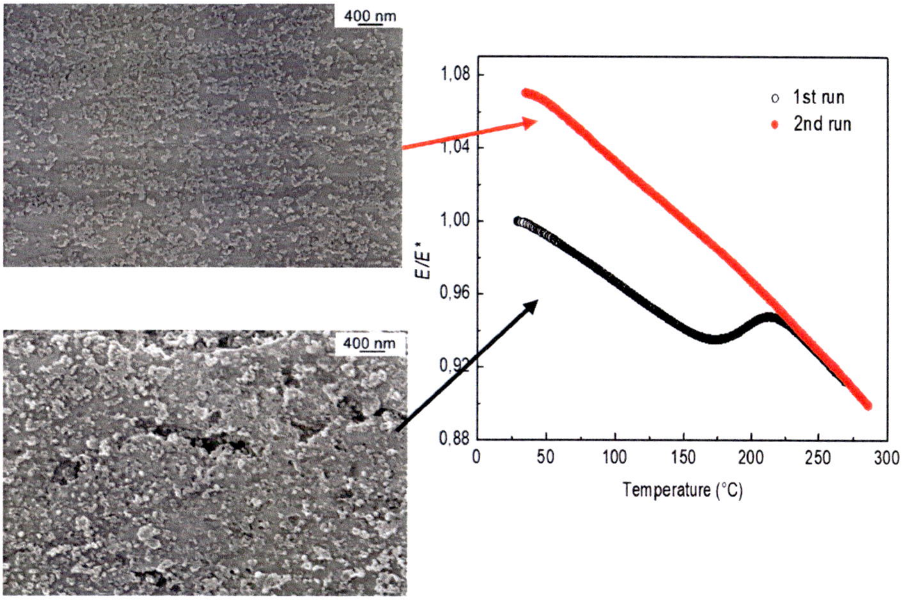

Fig. 7. Dynamic modulus E vs. temperature of AlSi10Mg manufactured by L-PBF measured in two successive test runs. The values of dynamic modulus are normalized to the value E* measured at room temperature in the 1st test run. The micrographs display pores in as-built material which are no more present after the 1st test run. Redrawn from Ref. [64].

In other cases heating leads to the closure of pores of nanometric size, as observed by some of present authors through measurements of dynamic modulus on AlSi10Mg manufactured by L-PBF [64, 65]. As shown in Fig. 7, dynamic modulus vs. temperature curve exhibits an inversion in the decreasing trend in the temperature range 170–210 °C. After cooling there is a permanent modulus variation of ~7%. The irreversible phenomenon, not observed in the successive test runs, is due to an increase of density connected to the closure of the smallest pores (nanometric size). The micrographs in Fig. 7 display pores in as-built material which are no more present after the 1st test run.

Owing to the high solidification rates, typical of AM processes, the as-printed Al alloys are often supersaturated solid solutions thus aging treatments can improve mechanical properties through the formation of precipitates [99, 100].

A significant example is represented by Sc-containing Al alloys developed first by soviet metallurgists and implemented on MiG-21 and MiG-29 aircrafts. Minor additions of Sc (<0.5%) remarkably increase the strength of Al alloys (>30%) because of Al3Sc precipitates homogeneously dispersed into the Al matrix. A second generation of these alloys is due to engineers at Airbus who added also Zr (Scalmalloy). Scalmalloy exhibits an excellent strength to weight ratio thus it is an ideal material for the aeronautic industry to use with TO tools. The alloy combines the lightness of Al with almost the same specific strength and ductility of Ti6Al4V [50, 101]. Moreover, it has a unique level of corrosion resistance and a stable microstructure also at elevated temperature.

Sc and Zr have been also used by Qi et al. [102] to develop a new Al-Li alloy (Al–Cu–Li–Sc–Zr) to be processed through L-PBF.

Another system of particular interest for high-temperature applications is the Al-Ce eutectic [103–112]. The strength at high temperature of the Al-Cu-Ce alloys depends on the spacing (~300 nm) and thermal stability of the intermetallics (Al8Cu3Ce) in the eutectic structure.

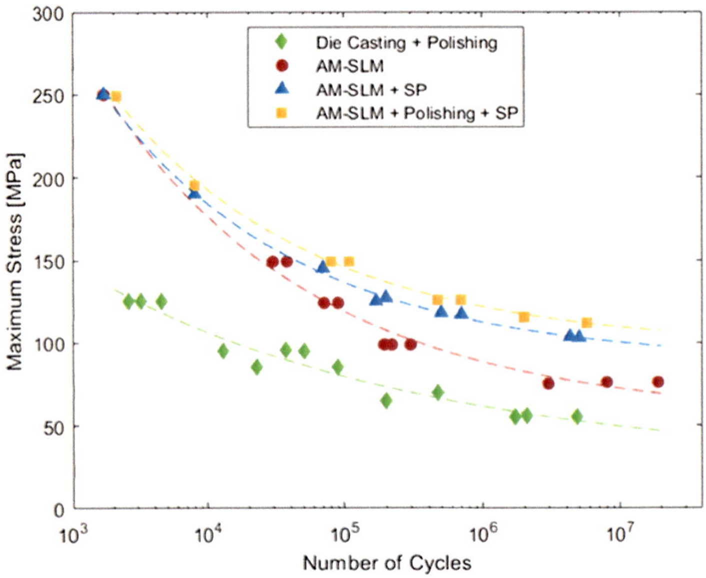

Fig. 8. S-N curves of die cast and L-PBF AlSi10Mg samples as printed, after polishing, and after polishing plus shot peening. (Redrawn from Ref. [92]).

Among metallic systems Al85Nd8Ni5Co2 alloy [113] produced by L-PBF has a composite microstructure with intermetallic AlNdNi4, Al4CoNi2 and AlNd3 platelets dispersed in the Al-matrix. Aversa et al. [114] improved the mechanical properties of AlSi10Mg alloy with the addition of Cu and Ag while Belelli et al. [115] modified the composition of the 2618 Al alloy with the addition of low amounts of Ti (2.3 wt.%) and B (0.8 wt.%).

Since AM components show lower fatigue resistance than conventionally produced ones, shot peening is one of the most effective treatments to improve this property. It induces a homogeneous residual stress state of compression that improves fatigue behavior [116]. Of particular relevance are the investigations of Uzan et al. [92] who found that the best results can be achieved by combining surface polishing and shot peening (Fig. 8).

4.2 Ti and Ti Alloys

To improve material properties solution heat treatment and aging [35] and annealing [117, 118] treatments can be performed on Ti components with the following results: solution heat treatment and aging lead to a columnar to equiaxed transition, and, with fast enough cooling rates, the formation of α' martensite [35]; annealing promotes a coarsening of the microstructure, prevents fatigue crack propagation [117] and improves creep response [118].

4.3 Mg and Mg Alloys

HIP treatment is useful to reduce the porosity also in the case of Mg alloys, however it has a significant drawback because the high temperature may alter the precipitate distribution. For instance, it has been demonstrated that the application of HIP on WE43 alloy printed by L-PBF promotes the dissolution of Nd-rich and Y-rich precipitates lowering the mechanical properties of the material [61].

The poor wear and corrosion resistance of Mg alloys limit their applications, especially in the aerospace industry. Conventional surface treatments, like plasma spraying of coatings of Al2O3-TiO2 on AZ31 Mg alloy, can be employed to improve these properties for AMed components as well [119].

5 Conclusions

The high standards required to components and materials and the strong competition in aerospace industry represent the driving force of recent advances in the use of AM technologies for processing light alloys The work provides an overview of current problems and possible solutions by focusing the attention on the strict relation between process parameters, microstructure and mechanical properties of Al, Ti and Mg alloys printed by various AM methods. The matter has been discussed by taking into account the aspects common to all the light alloys and those specific of each class of materials.

References

1. Vafadar, A., Guzzomi, F., Rassau, A., Hayward, K.: Advances in metal additive manufacturing: a review of common processes, industrial applications, and current challenges. Appl. Sci. **11**, 1213 (2021). https://doi.org/10.3390/app11031213
2. Gloria, A., Montanari, R., Richetta, M., Varone, A.: Alloys for aeronautic applications: state of the art and perspectives. Metals **9**, 662 (2019). https://doi.org/10.3390/met9060662
3. Blakey-Milner, B., et al.: Metal additive manufacturing in aerospace: a review. Mater. Des. **209**, 110008 (2021). https://doi.org/10.1016/j.matdes.2021.110008
4. Furuya, H., Kogiso, N., Matunaga, S., Senda, K.: Applications of magnesium alloys for aerospace structure systems. In: Magnesium Alloys. pp. 341–348. Trans Tech Publications Ltd (2000)
5. EOS webpage: https://www.eos.info/01_parts-and-applications/case_studies_applications_parts/_case_studies_pdf/en_cases/cs_m_aerospace_ruag_en.pdf. Last accessed Dec 2022
6. Altair, webpage: https://www.eos.info/01_parts-and-applications/case_studies_applications_parts/_case_studies_pdf/en_cases/cs_m_aerospace_ruag_en.pdf. Las accessed Dec 2022
7. 3D Systems, webpage: https://www.3dsystems.com/customer-stories/firstair-worthy-metal-printed-rf-filter-ready-take. Last accessed Dec 2022
8. Thales Alenia Space and 3D Systems, webpage: https://www.3dsystems.com/learningcenter/case-studies/tipping-point. Last accessed Dec 2022

9. Kellner, T.: 3D-Printed ‘Bionic’ Parts Could Revolutionize Aerospace Design. General Electric Reports. Boston, MA, USA (2017)
10. Huang, Y., et al.: Keyhole fluctuation and pore formation mechanisms during laser powder bed fusion additive manufacturing. Nat. Commun. **13**, 1170 (2022). https://doi.org/10.1038/s41467-022-28694-x
11. Weingarten, C., Buchbinder, D., Pirch, N., Meiners, W., Wissenbach, K., Poprawe, R.: Formation and reduction of hydrogen porosity during selective laser melting of AlSi10Mg. J. Mater. Process. Technol. **221**, 112–120 (2015). https://doi.org/10.1016/j.jmatprotec.2015.02.013
12. Zhang, Y., et al.: Additive manufacturing of metallic materials: a review. J. Mater. Eng. Perform. **27**(1), 1–13 (2017). https://doi.org/10.1007/s11665-017-2747-y
13. Ghio, E., Cerri, E.: Additive manufacturing of AlSi10Mg and Ti6Al4V lightweight alloys via laser powder bed fusion: a review of heat treatments effects. Materials. **15**, 2047 (2022). https://doi.org/10.3390/ma15062047
14. Hyer, H., Zhou, L., Benson, G., McWilliams, B., Cho, K., Sohn, Y.: Additive manufacturing of dense WE43 Mg alloy by laser powder bed fusion. Addit. Manuf. **33**, 101123 (2020). https://doi.org/10.1016/j.addma.2020.101123
15. Tang, M., Pistorius, P.C., Beuth, J.L.: Prediction of lack-of-fusion porosity for powder bed fusion. Addit. Manuf. **14**, 39–48 (2017). https://doi.org/10.1016/j.addma.2016.12.001
16. Read, N., Wang, W., Essa, K., Attallah, M.M.: Selective laser melting of AlSi10Mg alloy: process optimisation and mechanical properties development. Mater. Des. **1980–2015**(65), 417–424 (2015). https://doi.org/10.1016/j.matdes.2014.09.044
17. Cao, Y., Wei, H.L., Yang, T., Liu, T.T., Liao, W.H.: Printability assessment with porosity and solidification cracking susceptibilities for a high strength aluminum alloy during laser powder bed fusion. Addit. Manuf. **46**, 102103 (2021). https://doi.org/10.1016/j.addma.2021.102103
18. Stopyra, W., Gruber, K., Smolina, I., Kurzynowski, T., Kuźnicka, B.: Laser powder bed fusion of AA7075 alloy: influence of process parameters on porosity and hot cracking. Addit. Manuf. **35**, 101270 (2020). https://doi.org/10.1016/j.addma.2020.101270
19. Mishurova, T., Artzt, K., Haubrich, J., Requena, G., Bruno, G.: New aspects about the search for the most relevant parameters optimizing SLM materials. Addit. Manuf. **25**, 325–334 (2019). https://doi.org/10.1016/j.addma.2018.11.023
20. Li, Y., Gu, D.: Thermal behavior during selective laser melting of commercially pure titanium powder: numerical simulation and experimental study. Addit. Manuf. **1–4**, 99–109 (2014). https://doi.org/10.1016/j.addma.2014.09.001
21. Allavikutty, R., Gupta, P., Santra, T.S., Rengaswamy, J.: Additive manufacturing of Mg alloys for biomedical applications: current status and challenges. Curr. Opin. Biomed. Eng. **18**, 100276 (2021). https://doi.org/10.1016/j.cobme.2021.100276
22. Takagi, H., et al.: Material-property evaluation of magnesium alloys fabricated using wire-and-arc-based additive manufacturing. Addit. Manuf. **24**, 498–507 (2018). https://doi.org/10.1016/j.addma.2018.10.026
23. Tonelli, L., Liverani, E., Valli, G., Fortunato, A., Ceschini, L.: Effects of powders and process parameters on density and hardness of A357 aluminum alloy fabricated by selective laser melting. Int. J. Adv. Manuf. Technol. **106**(1–2), 371–383 (2019). https://doi.org/10.1007/s00170-019-04641-x
24. Poncelet, O., et al.: Critical assessment of the impact of process parameters on vertical roughness and hardness of thin walls of AlSi10Mg processed by laser powder bed fusion. Addit. Manuf. **38**, 101801 (2021). https://doi.org/10.1016/j.addma.2020.101801
25. Hrabe, N., Quinn, T.: Effects of processing on microstructure and mechanical properties of a titanium alloy (Ti–6Al–4V) fabricated using electron beam melting (EBM), Part 2: Energy

input, orientation, and location. Mater. Sci. Eng. A **573**, 271–277 (2013). https://doi.org/10.1016/j.msea.2013.02.065
26. Lee, Y., et al.: Effects of laser power on the microstructure evolution and mechanical properties of Ti–6Al–4V alloy manufactured by direct energy deposition. Met. Mater. Int. **28**(1), 197–204 (2021). https://doi.org/10.1007/s12540-021-01081-9
27. Promoppatum, P., Yao, S.-C.: Influence of scanning length and energy input on residual stress reduction in metal additive manufacturing: numerical and experimental studies. J. Manuf. Process. **49**, 247–259 (2020). https://doi.org/10.1016/j.jmapro.2019.11.020
28. Kempen, K., Thijs, L., Yasa, E., Badrossamay, M., Verheecke, W., Kruth, J.: Process optimization and microstructural analysis for selective laser melting of AlSi10Mg. In: 22nd Annual International Solid Freeform Fabrication Symposium—An Additive Manufacturing Conference (2011)
29. Galy, C., Le Guen, E., Lacoste, E., Arvieu, C.: Main defects observed in aluminum alloy parts produced by SLM: From causes to consequences. Addit. Manuf. **22**, 165–175 (2018). https://doi.org/10.1016/j.addma.2018.05.005
30. Kok, Y., et al.: Anisotropy and heterogeneity of microstructure and mechanical properties in metal additive manufacturing: a critical review. Mater. Des. **139**, 565–586 (2018). https://doi.org/10.1016/j.matdes.2017.11.021
31. Rashid, R., et al.: Effect of energy per layer on the anisotropy of selective laser melted AlSi12 aluminium alloy. Addit. Manuf. **22**, 426–439 (2018). https://doi.org/10.1016/j.addma.2018.05.040
32. Hönnige, J.R., Colegrove, P.A., Ganguly, S., Eimer, E., Kabra, S., Williams, S.: Control of residual stress and distortion in aluminium wire + arc additive manufacture with rolling. Addit. Manuf. **22**, 775–783 (2018). https://doi.org/10.1016/j.addma.2018.06.015
33. Wang, L., et al.: Investigation of performance and residual stress generation of AlSi10Mg processed by selective laser melting. Adv. Mater. Sci. Eng. **2018**, 7814039 (2018). https://doi.org/10.1155/2018/7814039
34. Buchbinder, D., Meiners, W., Pirch, N., Wissenbach, K., Schrage, J.: Investigation on reducing distortion by preheating during manufacture of aluminum components using selective laser melting. J. Laser Appl. **26**, 012004 (2014). https://doi.org/10.2351/1.4828755
35. Galarraga, H., Warren, R.J., Lados, D.A., Dehoff, R.R., Kirka, M.M., Nandwana, P.: Effects of heat treatments on microstructure and properties of Ti-6Al-4V ELI alloy fabricated by electron beam melting (EBM). Mater. Sci. Eng. A **685**, 417–428 (2017). https://doi.org/10.1016/j.msea.2017.01.019
36. Louvis, E., Fox, P., Sutcliffe, C.J.: Selective laser melting of aluminium components. J. Mater. Process. Technol. **211**, 275–284 (2011). https://doi.org/10.1016/j.jmatprotec.2010.09.019
37. Ullsperger, T., et al.: Ultra-short pulsed laser powder bed fusion of Al-Si alloys: impact of pulse duration and energy in comparison to continuous wave excitation. Addit. Manuf. **46**, 102085 (2021). https://doi.org/10.1016/j.addma.2021.102085
38. Li, S., et al.: Comparative study on the microstructures and properties of wire+arc additively manufactured 5356 aluminium alloy with argon and nitrogen as the shielding gas. Addit. Manuf. **34**, 101206 (2020). https://doi.org/10.1016/j.addma.2020.101206
39. Suryawanshi, J., Prashanth, K.G., Scudino, S., Eckert, J., Prakash, O., Ramamurty, U.: Simultaneous enhancements of strength and toughness in an Al-12Si alloy synthesized using selective laser melting. Acta Mater. **115**, 285–294 (2016). https://doi.org/10.1016/j.actamat.2016.06.009
40. Manfredi, D., Calignano, F., Krishnan, M., Canali, R., Ambrosio, E., Atzeni, E.: From powders to dense metal parts: characterization of a commercial AlSiMg alloy processed through direct metal laser sintering. Materials **6**, 856–869 (2013). https://doi.org/10.3390/ma6030856

41. Rosenthal, I., Stern, A., Frage, N.: Microstructure and mechanical properties of alsi10mg parts produced by the laser beam additive manufacturing (AM) technology. Metallogr. Microstruct. Anal. **3**(6), 448–453 (2014). https://doi.org/10.1007/s13632-014-0168-y
42. Chen, Y., Zhang, J., Gu, X., Dai, N., Qin, P., Zhang, L.-C.: Distinction of corrosion resistance of selective laser melted Al-12Si alloy on different planes. J. Alloy. Compd. **747**, 648–658 (2018). https://doi.org/10.1016/j.jallcom.2018.03.062
43. Kaufmann, N., Imran, M., Wischeropp, T.M., Emmelmann, C., Siddique, S., Walther, F.: Influence of process parameters on the quality of aluminium alloy EN AW 7075 using selective laser melting (SLM). Phys. Procedia **83**, 918–926 (2016). https://doi.org/10.1016/j.phpro.2016.08.096
44. Montero-Sistiaga, M.L., et al.: Changing the alloy composition of Al7075 for better processability by selective laser melting. J. Mater. Process. Technol. **238**, 437–445 (2016). https://doi.org/10.1016/j.jmatprotec.2016.08.003
45. Martin, J.H., Yahata, B.D., Hundley, J.M., Mayer, J.A., Schaedler, T.A., Pollock, T.M.: 3D printing of high-strength aluminium alloys. Nature **549**, 365–369 (2017). https://doi.org/10.1038/nature23894
46. Zhou, L., et al.: Microstructure and tensile property of a novel AlZnMgScZr alloy additively manufactured by gas atomization and laser powder bed fusion. Scripta Mater. **158**, 24–28 (2019). https://doi.org/10.1016/j.scriptamat.2018.08.025
47. Griffiths, S., et al.: Coarsening- and creep resistance of precipitation-strengthened Al–Mg–Zr alloys processed by selective laser melting. Acta Mater. **188**, 192–202 (2020). https://doi.org/10.1016/j.actamat.2020.02.008
48. Bi, J., et al.: Microstructure and mechanical properties of a novel Sc and Zr modified 7075 aluminum alloy prepared by selective laser melting. Mater. Sci. Eng. A **768**, 138478 (2019). https://doi.org/10.1016/j.msea.2019.138478
49. Zhou, L., et al.: Microstructure and mechanical properties of Zr-modified aluminum alloy 5083 manufactured by laser powder bed fusion. Addit. Manuf. **28**, 485–496 (2019). https://doi.org/10.1016/j.addma.2019.05.027
50. Spierings, A.B., et al.: Microstructural features of Sc- and Zr-modified Al-Mg alloys processed by selective laser melting. Mater. Des. **115**, 52–63 (2017). https://doi.org/10.1016/j.matdes.2016.11.040
51. Zhang, H., Zhu, H., Nie, X., Yin, J., Hu, Z., Zeng, X.: Effect of Zirconium addition on crack, microstructure and mechanical behavior of selective laser melted Al-Cu-Mg alloy. Scripta Mater. **134**, 6 (2017). https://doi.org/10.1016/j.scriptamat.2017.02.036
52. Hu, Z., Nie, X., Qi, Y., Zhang, H., Zhu, H.: Cracking criterion for high strength Al–Cu alloys fabricated by selective laser melting. Addit. Manuf. **37**, 101709 (2021). https://doi.org/10.1016/j.addma.2020.101709
53. Liu, S., Shin, Y.C.: Additive manufacturing of Ti6Al4V alloy: a review. Mater. Des. **164**, 107552 (2019). https://doi.org/10.1016/j.matdes.2018.107552
54. Ng, C.H., Bermingham, M.J., Dargusch, M.S.: Eliminating segregation defects during additive manufacturing of high strength β-titanium alloys. Addit. Manuf. **39**, 101855 (2021). https://doi.org/10.1016/j.addma.2021.101855
55. Tonelli, L.: Revealing the hierarchical microstructure of innovative additively manufactured metal parts with conventional light microscopy. Metallogr. Microstruct. Anal. **10**(2), 278–282 (2021). https://doi.org/10.1007/s13632-021-00721-1
56. Liu, X., Zhao, C., Zhou, X., Shen, Z., Liu, W.: Microstructure of selective laser melted AlSi10Mg alloy. Mater. Des. **168**, 107677 (2019). https://doi.org/10.1016/j.matdes.2019.107677
57. Eskandari, H., Lashgari, H.R., Zangeneh, Sh., Kong, C., Ye, L., Eizadjou, M., Wang, H.: Microstructural characterization and mechanical properties of SLM-printed Ti–6Al–4V

alloy: effect of build orientation. J. Mater. Res. **37**, 2645–2660 (2022). https://doi.org/10.1557/s43578-021-00468-z
58. Bär, F., et al.: Laser additive manufacturing of biodegradable magnesium alloy WE43: a detailed microstructure analysis. Acta Biomater. **98**, 36–49 (2019). https://doi.org/10.1016/j.actbio.2019.05.056
59. Al-Bermani, S.S., Blackmore, M.L., Zhang, W., Todd, I.: The origin of microstructural diversity, texture, and mechanical properties in electron beam melted Ti-6Al-4V. Metall. and Mater. Trans. A. **41**, 3422–3434 (2010). https://doi.org/10.1007/s11661-010-0397-x
60. Bassani, P., Biffi, C.A., Casati, R., Zanatta Alarcon, A., Tuissi, A., Vedani, M.: Properties of aluminium alloys produced by selective laser melting. Key Eng. Mater. **710**, 83–88 (2016). https://doi.org/10.4028/www.scientific.net/KEM.710.83
61. Gangireddy, S., Gwalani, B., Liu, K., Faierson, E.J., Mishra, R.S.: Microstructure and mechanical behavior of an additive manufactured (AM) WE43-Mg alloy. Addit. Manuf. **26**, 53–64 (2019). https://doi.org/10.1016/j.addma.2018.12.015
62. Manakari, V., Parande, G., Gupta, M.: Selective laser melting of magnesium and magnesium alloy powders: a review. Metals. **7**, 2 (2016). https://doi.org/10.3390/met7010002
63. Wei, K., Gao, M., Wang, Z., Zeng, X.: Effect of energy input on formability, microstructure and mechanical properties of selective laser melted AZ91D magnesium alloy. Mater. Sci. Eng. A **611**, 212–222 (2014). https://doi.org/10.1016/j.msea.2014.05.092
64. Cabibbo, M., Montanari, R., Pola, A., Tocci, M., Varone, A.: Mechanical spectroscopy study of as-cast and additive manufactured AlSi10Mg. J. Alloy. Compd. **914**, 165361 (2022). https://doi.org/10.1016/j.jallcom.2022.165361
65. Tocci, M., Varone, A., Montanari, R., Pola, A.: Study of high temperature properties of alsi10mg alloy produced by laser-based powder bed fusion. In: THERMEC, pp. 1485–1491. Trans Tech Publications Ltd (2021)
66. Tang, M., Pistorius, P.C.: Oxides, porosity and fatigue performance of AlSi10Mg parts produced by selective laser melting. Int. J. Fatigue **94**, 192–201 (2017). https://doi.org/10.1016/j.ijfatigue.2016.06.002
67. Nicoletto, G., Konečná, R., Frkáň, M., Riva, E.: Surface roughness and directional fatigue behavior of as-built EBM and DMLS Ti6Al4V. Int. J. Fatigue **116**, 140–148 (2018). https://doi.org/10.1016/j.ijfatigue.2018.06.011
68. Yang, K.V., Shi, Y., Palm, F., Wu, X., Rometsch, P.: Columnar to equiaxed transition in Al-Mg(-Sc)-Zr alloys produced by selective laser melting. Scripta Mater. **145**, 113–117 (2018). https://doi.org/10.1016/j.scriptamat.2017.10.021
69. Li, X.P., et al.: Selective laser melting of nano-TiB2 decorated AlSi10Mg alloy with high fracture strength and ductility. Acta Mater. **129**, 183–193 (2017). https://doi.org/10.1016/j.actamat.2017.02.062
70. Rao, H., Giet, S., Yang, K., Wu, X., Davies, C.H.J.: The influence of processing parameters on aluminium alloy A357 manufactured by selective laser melting. Mater. Des. **109**, 334–346 (2016). https://doi.org/10.1016/j.matdes.2016.07.009
71. Liu, Y.J., Liu, Z., Jiang, Y., Wang, G.W., Yang, Y., Zhang, L.C.: Gradient in microstructure and mechanical property of selective laser melted AlSi10Mg. J. Alloy. Compd. **735**, 1414–1421 (2018). https://doi.org/10.1016/j.jallcom.2017.11.020
72. Hadadzadeh, A., Shalchi Amirkhiz, B., Odeshi, A., Li, J., Mohammadi, M.: Role of hierarchical microstructure of additively manufactured AlSi10Mg on dynamic loading behavior. Addit. Manuf. **28**, 1–13 (2019). https://doi.org/10.1016/j.addma.2019.04.012
73. Xiong, Z.H., Liu, S.L., Li, S.F., Shi, Y., Yang, Y.F., Misra, R.D.K.: Role of melt pool boundary condition in determining the mechanical properties of selective laser melting AlSi10Mg alloy. Mater. Sci. Eng. A **740–741**, 148–156 (2019). https://doi.org/10.1016/j.msea.2018.10.083

74. Hadadzadeh, A., Amirkhiz, B.S., Langelier, B., Li, J., Mohammadi, M.: Microstructural consistency in the additive manufactured metallic materials: a study on the laser powder bed fusion of AlSi10Mg. Addit. Manuf. **46**, 102166 (2021). https://doi.org/10.1016/j.addma.2021.102166
75. Hadadzadeh, A., Amirkhiz, B.S., Li, J., Mohammadi, M.: Columnar to equiaxed transition during direct metal laser sintering of AlSi10Mg alloy: effect of building direction. Addit. Manuf. **23**, 121–131 (2018). https://doi.org/10.1016/j.addma.2018.08.001
76. Maizza, G., Pero, R., Richetta, M., Montanari, R.: Continuous dynamic recrystallization (CDRX) model for aluminum alloys. J. Mater. Sci. **53**(6), 4563–4573 (2017). https://doi.org/10.1007/s10853-017-1845-4
77. Tedde, G.M., di Schino, A., Donnini, R., Montanari, R., Richetta, M., Santo, L., Testani, C., Varone, A.: An innovative industrial process for forging 7050 Al alloy. In: THERMEC, pp. 1047–1052. Trans Tech Publications Ltd (2019)
78. Angella, G., Di Schino, A., Donnini, R., Richetta, M., Testani, C., Varone, A.: AA7050 Al alloy hot-forging process for improved fracture toughness properties. Metals. **9**, 64 (2019). https://doi.org/10.3390/met9010064
79. Di Schino, A., Montanari, R., Testani, C., Varone, A.: Dislocation breakaway damping in AA7050 alloy. Metals **10**, 1682 (2020). https://doi.org/10.3390/met10121682
80. Altıparmak, S.C., Yardley, V.A., Shi, Z., Lin, J.: Challenges in additive manufacturing of high-strength aluminium alloys and current developments in hybrid additive manufacturing. Int. J. Light. Mater. Manuf. **4**, 246–261 (2021). https://doi.org/10.1016/j.ijlmm.2020.12.004
81. Hauser, T., et al.: Oxidation in wire arc additive manufacturing of aluminium alloys. Addit. Manuf. **41**, 101958 (2021). https://doi.org/10.1016/j.addma.2021.101958
82. Herzog, D., Bartsch, K., Bossen, B.: Productivity optimization of laser powder bed fusion by hot isostatic pressing. Addit. Manuf. **36**, 101494 (2020). https://doi.org/10.1016/j.addma.2020.101494
83. Kasperovich, G., Hausmann, J.: Improvement of fatigue resistance and ductility of TiAl6V4 processed by selective laser melting. J. Mater. Process. Technol. **220**, 202–214 (2015). https://doi.org/10.1016/j.jmatprotec.2015.01.025
84. Tan, X., et al.: Graded microstructure and mechanical properties of additive manufactured Ti–6Al–4V via electron beam melting. Acta Mater. **97**, 1–16 (2015). https://doi.org/10.1016/j.actamat.2015.06.036
85. Carroll, B.E., Palmer, T.A., Beese, A.M.: Anisotropic tensile behavior of Ti–6Al–4V components fabricated with directed energy deposition additive manufacturing. Acta Mater. **87**, 309–320 (2015). https://doi.org/10.1016/j.actamat.2014.12.054
86. Gockel, J., Beuth, J., Taminger, K.: Integrated control of solidification microstructure and melt pool dimensions in electron beam wire feed additive manufacturing of Ti-6Al-4V. Addit. Manuf. **1–4**, 119–126 (2014). https://doi.org/10.1016/j.addma.2014.09.004
87. Alaghmandfard, R., Chalasani, D., Hadadzadeh, A., Amirkhiz, B.S., Odeshi, A., Mohammadi, M.: Dynamic compressive response of electron beam melted Ti–6Al–4V under elevated strain rates: microstructure and constitutive models. Addit. Manuf. **35**, 101347 (2020). https://doi.org/10.1016/j.addma.2020.101347
88. Zumdick, N.A., Jauer, L., Kersting, L.C., Kutz, T.N., Schleifenbaum, J.H., Zander, D.: Additive manufactured WE43 magnesium: a comparative study of the microstructure and mechanical properties with those of powder extruded and as-cast WE43. Mater. Charact. **147**, 384–397 (2019). https://doi.org/10.1016/j.matchar.2018.11.011
89. Karunakaran, R., Ortgies, S., Tamayol, A., Bobaru, F., Sealy, M.P.: Additive manufacturing of magnesium alloys. Bioact. Mater. **5**, 44–54 (2020). https://doi.org/10.1016/j.bioactmat.2019.12.004

90. Gieseke, M., Noelke, C., Kaierle, S., Wesling, V., Haferkamp, H.: Selective laser melting of magnesium and magnesium alloys. In: Hort, N., Mathaudhu, S.N., Neelameggham, N.R., Alderman, M. (eds.) Magnesium Technology, pp. 65–68. Springer, Cham (2013). https://doi.org/10.1007/978-3-319-48150-0_11
91. Guo, Y., Pan, H., Ren, L., Quan, G.: Microstructure and mechanical properties of wire arc additively manufactured AZ80M magnesium alloy. Mater. Lett. **247**, 4–6 (2019). https://doi.org/10.1016/j.matlet.2019.03.063
92. Uzan, N.E., Ramati, S., Shneck, R., Frage, N., Yeheskel, O.: On the effect of shot-peening on fatigue resistance of AlSi10Mg specimens fabricated by additive manufacturing using selective laser melting (AM-SLM). Addit. Manuf. **21**, 458–464 (2018). https://doi.org/10.1016/j.addma.2018.03.030
93. Uzan, N.E., Shneck, R., Yeheskel, O., Frage, N.: Fatigue of AlSi10Mg specimens fabricated by additive manufacturing selective laser melting (AM-SLM). Mater. Sci. Eng., A **704**, 229–237 (2017). https://doi.org/10.1016/j.msea.2017.08.027
94. Aliprandi, P., Giudice, F., Guglielmino, E., Sili, A.: Tensile and creep properties improvement of Ti-6Al-4V alloy specimens produced by electron beam powder bed fusion additive manufacturing. Metals **9**, 1207 (2019). https://doi.org/10.3390/met9111207
95. Hu, Z., et al.: Aging responses of an Al-Cu alloy fabricated by selective laser melting. Addit. Manuf. **37**, 101635 (2021). https://doi.org/10.1016/j.addma.2020.101635
96. Maamoun, A.H., Elbestawi, M., Dosbaeva, G.K., Veldhuis, S.C.: Thermal post-processing of AlSi10Mg parts produced by selective laser melting using recycled powder. Addit. Manuf. **21**, 234–247 (2018). https://doi.org/10.1016/j.addma.2018.03.014
97. Cerri, E., Ghio, E., Bolelli, G.: Defect-correlated vickers microhardness of Al-Si-Mg alloy manufactured by laser powder bed fusion with post-process heat treatments. J. Mater. Eng. Perform. **31**, 8047–8067 (2022). https://doi.org/10.1007/s11665-022-06874-6
98. Giovagnoli, M., et al.: Effect of different heat-treatment routes on the impact properties of an additively manufactured AlSi10Mg alloy. Mater. Sci. Eng. A **802**, 140671 (2021). https://doi.org/10.1016/j.msea.2020.140671
99. Casati, R., Coduri, M., Riccio, M., Rizzi, A., Vedani, M.: Development of a high strength Al–Zn–Si–Mg–Cu alloy for selective laser melting. J. Alloy. Compd. **801**, 243–253 (2019). https://doi.org/10.1016/j.jallcom.2019.06.123
100. Fite, J., Prameela, S.E., Slotwinski, J.A., Weihs, T.P.: Evolution of the microstructure and mechanical properties of additively manufactured AlSi10Mg during room temperature holds and low temperature aging. Addit. Manuf. **36**, 101429 (2020). https://doi.org/10.1016/j.addma.2020.101429
101. Jia, Q., Rometsch, P., Kürnsteiner, P., Chao, Q., Huang, A., Weyland, M., Bourgeois, L., Wu, X.: Selective laser melting of a high strength Al-Mn-Sc alloy: alloy design and strengthening mechanisms. Acta Mater. **171** (2019). https://doi.org/10.1016/j.actamat.2019.04.014
102. Qi, Y., Hu, Z., Zhang, H., Nie, X., Zhang, C., Zhu, H.: High strength Al–Li alloy development for laser powder bed fusion. Addit. Manuf. **47**, 102249 (2021). https://doi.org/10.1016/j.addma.2021.102249
103. Bahl, S., et al.: Al-Cu-Ce(-Zr) alloys with an exceptional combination of additive processability and mechanical properties. Addit. Manuf. **48**, 102404 (2021). https://doi.org/10.1016/j.addma.2021.102404
104. Czerwinski, F.: Cerium in aluminum alloys. J. Mater. Sci. **55**(1), 24–72 (2019). https://doi.org/10.1007/s10853-019-03892-z
105. Czerwinski, F.: Thermal stability of aluminum Alloys. Mater. **13**, 3441 (2020). https://doi.org/10.3390/ma13153441
106. Czerwinski, F.: Critical assessment 40: a search for the eutectic system of high-temperature cast aluminium alloys. Mater. Sci. Technol. **37**, 683–692 (2021). https://doi.org/10.1080/02670836.2021.1940670

107. Czerwinski, F.: Thermal stability of aluminum–cerium binary alloys containing the Al–Al11Ce3 eutectic. Mater. Sci. Eng. A **809**, 140973 (2021). https://doi.org/10.1016/j.msea.2021.140973
108. Ng, D., Dunand, D.: Aging- and creep-resistance of a cast hypoeutectic Al-6.9Ce-9.3Mg (wt.%) alloy. Mater. Sci. Eng.: A. **786**, 139398 (2020). https://doi.org/10.1016/j.msea.2020.139398
109. Plotkowski, A., et al.: Microstructure and properties of a high temperature Al–Ce–Mn alloy produced by additive manufacturing. Acta Mater. **196**, 595–608 (2020). https://doi.org/10.1016/j.actamat.2020.07.014
110. Yang, Y., et al.: Primary solidification of ternary compounds in Al-rich Al–Ce–Mn alloys. J. Alloy. Compd. **844**, 156048 (2020). https://doi.org/10.1016/j.jallcom.2020.156048
111. Zhou, L., Huynh, T., Park, S., Hyer, H., Mehta, A., Song, S., Bai, Y.: Laser powder bed fusion of Al-10 wt.% Ce alloys: microstructure and tensile property. J. Mater. Sci. **55**, 14611+ (2020)
112. Dantzig, J.A., Rappaz, M.: Solidification: Revised and Expanded. EPFL press (2016)
113. Prashanth, K.G., et al.: Production of high strength Al85Nd8Ni5Co2 alloy by selective laser melting. Addit. Manuf. **6**, 1–5 (2015). https://doi.org/10.1016/j.addma.2015.01.001
114. Aversa, A., Bosio, F., Marola, S., Lorusso, M., Manfredi, D., Battezzati, L., Fino, P., Lombardi, M.: Laser single scan tracks of new aluminium alloys compositions. In: Proceedings of EuroPM 2018 Congress and Exhibition, pp. 15–20. European Powder Metallurgy Association (2020)
115. Belelli, F., Casati, R., Larini, F., Riccio, M., Vedani, M.: Investigation on two Ti–B-reinforced Al alloys for laser powder bed fusion. Mater. Sci. Eng. A **808**, 140944 (2021). https://doi.org/10.1016/j.msea.2021.140944
116. Salmi, A., Atzeni, E., Iuliano, L., Galati, M.: Experimental analysis of residual stresses on AlSi10Mg parts produced by means of selective laser melting (SLM). Procedia CIRP **62**, 458–463 (2017). https://doi.org/10.1016/j.procir.2016.06.030
117. Kim, S., Oh, H., Kim, J.G., Kim, S.: Effect of annealing and crack orientation on fatigue crack propagation of Ti64 alloy fabricated by direct energy deposition process. Met. Mater. Int. **28**(1), 205–215 (2021). https://doi.org/10.1007/s12540-021-01087-3
118. Paoletti, C., Cabibbo, M., Santecchia, E., Cerri, E., Spigarelli, S.: Effect of post-processing heat treatments on short-term creep response at 650 °C for a Ti-6Al-4V alloy produced by additive manufacturing. Metals **12**, 1084 (2022). https://doi.org/10.3390/met12071084
119. Kurzynowski, T., Pawlak, A., Smolina, I.: The potential of SLM technology for processing magnesium alloys in aerospace industry. Arch. Civ. Mech. Eng. **20**(1), 1–13 (2020). https://doi.org/10.1007/s43452-020-00033-1

Continuum Robots for Space Applications

Matteo Russo(✉)

Department of Industrial Engineering, University of Rome Tor Vergata, 00133 Roma, Italy
matteo.russo@uniroma2.it

Abstract. Continuum robots are flexible structures whose backbone shape can be actively controlled through actuators such as tendons and fluid chambers. Often characterized by a slender design with a high length to cross-section diameter ratio enables operation in hard-to-reach environments with narrow and tortuous passages. This makes them suited for inspection tasks in a large variety of industrial settings, including spacecraft. Further, their high redundancy enables dexterous maintenance tasks such as satellite repair. In this paper, current and potential applications of continuum robots in space are discussed with regards to the needs and requirements of space robotics, with a perspective on open challenges and future research directions.

Keywords: Continuum robots · Soft robotics · Space · Orbital robotics · Aerospace

1 Introduction

Continuum robots are continuously bending structures whose shape can be actively controlled through actuators such as tendons, pneumatics, and smart materials [1]. Bioinspired by trunks, tendrils, and snakes, the flexible bodies of these robots can be designed with high length to cross-section diameter ratio. This makes them uniquely suited to navigating narrow, tortuous paths and deploying instrumentation in hard-to-reach environments. While research on continuum robots often focuses on medical applications [2], several industrial use cases have been recently proposed, including in aerospace. In this paper, the potential of continuum robots for space operation is discussed, with a review of existing space-related designs and a perspective on open challenges and future research directions.

1.1 Space Robotics

Space robotics covers both planetary and orbital scenarios [3] (Fig. 1). Planetary robotics aims at surface exploration and instrument deployment [4]. Orbital robotics focuses on service activities on spacecraft, providing satellites, shuttles, and space stations with inspection, maintenance, and assembly capabilities [5]. From a mechanical perspective, planetary and orbital robots require fundamentally different designs, as surface exploration favors rovers and other structures with ground mobility, whereas orbital operation

© The Author(s), under exclusive license to Springer Nature Switzerland AG 2023
M. Ceccarelli et al. (Eds.): TORVEASTRO 2023, MMS 130, pp. 129–139, 2023.
https://doi.org/10.1007/978-3-031-28447-2_10

usually requires in-space manipulation [4, 5]. As continuum robots are particularly suited to inspection and maintenance in hard-to-reach areas, orbital scenarios are naturally more interesting for them. Nevertheless, some planetary applications for rover assistance have also been proposed.

Fig. 1. Examples of planetary and orbital space robotics: (a) Planetary application with the Mars Rover (NASA, [8]); (b) Orbital application with the *Canadarm* (ESA, [5]).

Robotic systems for both planetary and orbital exploration show various degrees of autonomy. Generally, planetary rovers require full autonomy, whereas orbital systems can also be semi-autonomous or teleoperated [6]. In this, the main challenge is to achieve the robustness and full system integration required by such challenging scenarios [7]. Whether teleoperated or autonomous, a space robot is expected to function unhindered and with minimal maintenance for the longest time possible. For example, the NASA Mars Exploration Rovers *Spirit* and *Opportunity*, both designed for a 90-day mission, lived up to 6 and 14 years respectively (Fig. 1a) [8]. Their successor *Curiosity*, with an expected lifespan of approximately 2 years, has also been roaming Mars for over

10 years [9]. Similar robustness and reliability have been shown by orbital robots such as the *Mobile Servicing System* (MSS) of the International Space Station (ISS, Fig. 1b), launched in 2001 and still operational after more than 20 years [5].

The need for robustness deeply affects the development of a space robot. For example, uncertainties on kinematic and dynamic model parameters and general manufacturing and assembly tolerances will negatively affect the robot's lifespan. Redundant, fault-tolerant designs are needed to keep operating after accidents or adversities. These aspects are of foremost importance for continuum robots, as their flexible bodies can increase uncertainty.

Another key challenge of space operations is microgravity. Conventional robots are design to function under Earth-level gravity, and a reduced gravity compromises the functioning of systems that we usually take for granted, such as wheels [10]. In-space operation also requires motion modes that are never observed on Earth. For example, free-flying in proximity of large orbital structures requires specific collision-free trajectory planning [11–13]. Further, when operating on a spacecraft, robot motion causes a disturbance to the spacecraft itself [5]. From a control perspective, microgravity must be carefully modeled to predict and minimize the consequences of robot motion onto spacecraft and related systems [14].

Robustness and microgravity represent the foremost challenges of general space robotics, but many other issues can arise for specific space robots and use cases [15–18]. This paper highlights and discusses the ones related to continuum robots.

1.2 Continuum Robots

In conventional robots, motion only happens at joints, while most of the structure, made of stiff bodies (often assumed as rigid), undergoes negligible deformation. Conversely, the entire body of a continuum robot deforms under load. This feature makes continuum robots more flexible and adaptable than rigid-body manipulators. Thanks to their theoretically infinite degrees of freedom, they show a better performance in unstructured or extremely complex environments. However, this comes at a cost: where a rigid-body robot is described by a limited set of parameters and variables, an exact continuum model requires a continuous differential. Further, rigid-body kinematics can be abstracted from rigid-body dynamics, and the assumption of rigid body makes material mechanics superfluous. In a continuum model, kinematics, dynamics, and material mechanics are instead strictly intertwined. Model and control complexity thus increase, resulting in a less efficient and accurate system.

A strong research effort has been placed on exploiting the benefits of continuum structures while minimizing these drawbacks. Simplified models, such as piecewise constant curvature (PCC) kinematics [19], have been proposed to reduce the order of the system (i.e., approximate its configuration with a finite number of motion parameters) [20–23]. A learning-based, model-free approach can also help address this challenge [24]. Specific, more efficient approaches work for continuum robot subfamilies, such as low-compliance *soft robots*, ideal for unstructured environments [25–27] (Fig. 2a), and miniaturized *concentric tube robots*, with an extremely small diameter (less than 2 mm) but shorter reach [28, 29].

Most application-driven research on continuum robot is aimed at the medical field, as demonstrated by the large number of surveys on the topic [2, 30–33]. The research effort behind industrial applications (including space) has been more scattered for a variety of reasons, such as a different scale, higher costs, the large range of use cases, industrial or military interests preventing disclosure, and more. These issues have not prevented the publications of a rich – even if fragmented – literature on industrial continuum robots, with many examples for operation in nuclear facilities [34–36], on aircrafts [37–41] (Fig. 2b), and in space. The latter is the focus of this survey, which aims at collecting and classifying previous research works on continuum robots in space.

a b

Fig. 2. Examples of continuum robot architectures: (a) A soft growing continuum robot (Stanford University, credit: L.A. Cicero [27]); (b) A tendon-driven continuum robot for aerospace (University of Nottingham and Rolls-Royce, [36–38]).

2 State of the Art

While no previous survey discusses continuum robots for space applications only, two previous works on soft robots (which can be seen as a subfamily of continuum robots) address the topic. In [42], among a general overview on soft robotics applications, the authors highlight the capability of continuum robots to inspect hard-to-reach remote areas such as the crevasses and thermal blankets covering instrumentation on spacecraft. The survey in [43] conversely, only discusses space applications of soft robots but within a limited design range: the authors only consider fluid and smart material actuation of soft materials, there defined as those with Young's modulus lower than 1 GPa. As such, they identify material durability as the main limitations of these robots in space, as they can hardly provide the required reliability. However, those designs only represent a small subgroup of all continuum robots, which can be made with more reliable technologies (precision manufacturing) and materials (superelastic nickel-titanium alloys, i.e., nitinol). As such, a partially different set of challenges that is distinct from the one in [43] is faced when developing space continuum robots, as highlighted by the examples below.

The first continuum robots demonstrated in space applications is arguably the *NASA tendril robot*, first presented in [44]. With a cross-section diameter to length ration of 1:100 (10 mm–1 m), the NASA tendril robot has a camera end-effector for visual inspection. This early design uses helicoidal springs as compliant/continuum elements. These elements are actuated by pulling antagonistic tendons from the base of the robot, for a total mobility of four degrees of freedom (divided among the two independent 2-DoF segments).

A rich literature of continuum robots for space applications has been built by Prof. Ian Walker's research group at Clemson University, as seen in [45–52]. Mainly addressing the inspection of spacecraft (and, in particular, of the ISS), their research focuses on new design features and modeling improvements to enhance performance. The 3-segment continuum robot in [45, 46] is based on spring-loaded concentric carbon fiber tubes and is 1830 mm long, with a 1:130 cross-section diameter to length ratio and the capability to "compress" its length by up to 40%. The main novel design characteristic of this robot is the addition of hooks along the backbone to passively grab environmental features [45]. A further development is reported in [47], where a segment design that stiffens by layer jamming is introduced. While improving performance, each stiffening segment is 415 mm long with an external diameter of 63 mm. This leads to an overall increase in diameter and a reduction in slenderness (1:20 cross-section diameter to length ratio for a 3-segment robot). Demonstrations of these robots in artificial vegetation and an ISS mockup are shown in [48] and [49], respectively. Model-less control strategies are described in [50] and [51]. Finally, a different application for these robots is proposed in [52], where the usage of "active hose" for pumping viscous fluids is explored.

The research from Clemson University focuses on tendon-driven extensible continuum robots for inspection tasks, but different research groups have suggested alternative uses and architectures. For example, deployable antennas are among the proposed applications for the soft growing technology introduced in [53]. In [54], the tip-extending and potentially forking body makes these soft growing robots capable of acting as antennas. Space exploration scenarios are mentioned as one of the two core applications (the other being search and rescue), even if no demonstration is reported.

On-orbit satellite surgery is the focus of the work in [55], where a combination of two rigid-link manipulators with 7 degrees of freedom each and two continuum "catheter arms" is shown to successfully operate on a satellite mockup. The two continuum arms here have a mostly passive role, rigidizing once their end-effector is brought to the right position by the active rigid-link manipulators. A different preliminary work with an active tendon-driven prototype for satellite surgery is presented in [56], with a spring-loaded backbone.

Overall, the main features of these space continuum robots are summarized in Table 1. By analyzing this literature, the key feature of continuum robot for space applications is their slenderness (i.e., diameter to length ratio). While many continuum robots for conventional applications are characterized by a strict constraint on either diameter or length, outer space and planetary exploration favor a more performing form factor. From an architectural point of view, spacecraft and planetary exploration differ in necessities: spacecraft require a higher degree of accuracy, and thus spring-based robots are preferred in this scenario; conversely, soft growing robots can better adapt to the unstructured environments of planetary (or asteroid) surfaces.

However, this literature also highlights how no continuum robot was tested in space, and how much the technology must evolve before reaching the robustness required: the high positioning uncertainty of continuum robots and the limited testing made on long-term performance and fatigue life of the compliant or soft components clashes with the reliability demand of space robotics.

Table 1. Examples of space continuum robotics.

Refs.	Applications	Actuation	Backbone	DoF	Length (L)	Diameter (D)	D:L ratio	Features
[44]	Spacecraft inspection	Tendons	Spring-loaded concentric tubes	4	1.0 m	10 mm	1:100	–
[45]	Spacecraft inspection	Tendons	Spring-loaded concentric tubes	6	1.8 m	–	1:130	Hooks
[47]	Spacecraft inspection	Tendons	Spring-loaded concentric tubes	6	1.2 m	63 mm	1:20	Stiffening
[54]	Deployable antennas	Fluidic	Soft	–	Varies	Varies	–	Growing, adaptive
[55]	Satellite surgery	–	Passive tube	–	–	–	–	Stiffening
[56]	Satellite surgery	Tendons	Springs	6	–	–	–	–

3 Perspective

3.1 Robustness

As concluded in the previous section, the key challenge of all continuum robots, and arguably not only space-oriented ones, is their lack of robustness. Because of system complexity and intrinsic compliance, an accurate prediction of continuum robot behavior is almost impossible to achieve. Furthermore, most continuum robots have been only tested at a prototype stage. Thus, no study analyzes performance deterioration with time, as fatigue cycles could cause a change in mechanical properties (or, in a worst case, failure) of their compliant elements.

A related issue, especially for non-machinable structures (e.g., with soft material rather than superelastic alloys as compliant elements), arises from fabrication. Many soft robots are currently manufactured in an artisanal or semi-artisanal way rather than by following a strict and repeatable procedure that can lead to industrialization and, later, industry uptake. While space applications do not require neither mass production nor large production numbers, a repeatable procedure that leads to reliable and homogeneous components is still critical for development and testing.

Fault management also needs to be considered. What happens if a tendon of a tendon-driven robot snaps? What if the pressurized chamber of a soft growing robot is pierced? Damage that can be repaired easily on Earth might be impossible to repair in space and lead to much worse consequences. As such, this issue needs to be addressed in advance with both fault-tolerant designs and control systems (for example, by borrowing from research on self-tuning and failure management from similar designs [57, 58]).

3.2 Microgravity

Some elements of continuum robots might be affected by microgravity. For example, pneumatic and fluid-based actuation will act differently with the change in pressure, and both design and controllers need to consider this for successful operation.

However, out of all the challenges that robots face in space, reduced gravity is the most likely to benefit, rather than hinder, continuum robots. The length of continuum robots that operate in "suspended" configurations (i.e., without unloading part of their weight onto the environment) is generally bounded by their tip displacement under gravitational load, due to their low stiffness. In space, microgravity could thus enable longer, more accurate, and more dynamically performing continuum robots.

3.3 Environmental Hazards

Space robots face two further hazards that are rarely encountered on Earth, which particularly affect their electronics: radiation and temperature.

Systems in low orbit and outer space are affected by a ionizing radiation due to a combination of sources, including nuclear fusion in the Sun and other stars, particles trapped by planetary electromagnetic fields (e.g. Earth's Van Allen radiation belt), galactic cosmic rays, and secondary x-rays [59]. This radiation degrades electrical and electronic components, quickly causing critical failures. Thus, radiation hardened components (with the associated costs and design complications) must be used to ensure robot functioning in space. As conventional robots (i.e. rigid-link manipulators) distribute motors throughout the robot's body, directly connecting them to the joints, each part of the robot needs to be radiation hardened to ensure its functioning. Conversely, continuum robots (especially tendon-driven and pneumatically actuated ones) can be designed with an external actuation pack at its base, containing the entire control hardware (including all the electronics) and transmitting motions through tendons and fluid only to the robot's body. As such, only the actuation pack requires radiation hardening, significantly reducing costs (both economical and from a complexity/performance perspective) and simplifying inspection and maintenance.

The environmental temperature in low orbit ranges from approximately 210 to 400 K. Thus, hardware needs to be designed to function in this extreme range. From an electrical and electronic perspective, this can be achieved with a proper insulation of the actuation pack. However, differently from radiation, the effect of temperature can be critical on mechanical components as well. For example, many continuum robots for aerospace employ nitinol (Nickel-Titanium alloys) as backbone material. This alloy is super-elastic and characterized by a martensite to austenite transformation when under load and with variations of temperature; as such, designers using similar materials for continuum robots in space need to consider not only how system dynamics affect material properties, but also how material properties vary with temperature. Similar considerations must be done for the selection of other materials.

3.4 Task-Oriented Designs

A major gap in current literature is represented by continuum robots specifically designed for space operation. The robots in the reported literature are arguably applications of conventional continuum designs to space applications. While a space scenario is proposed for all of them, the robot development is then approached in a conventional way, without discussing the challenges created by spacecraft or planetary exploration. As such, a more in-depth analysis of those challenges, how they affect continuum designs, and potential solutions should be discussed in future works on the topic.

References

1. Walker, I.D., Choset, H., Chirikjian, G.S., Snake-like and continuum robots. Springer handbook of robotics, pp. 481–498. Springer, Cham (2016)
2. Burgner-Kahrs, J., Rucker, D.C., Choset, H.: Continuum robots for medical applications: a survey. IEEE Trans. Rob. **31**(6), 1261–1280 (2015)
3. Pedersen, L., Kortenkamp, D., Wettergreen, D., Nourbakhsh, I., Korsmeyer, D.: A survey of space robotics. In: ISAIRAS (2003)
4. Yoshida, K.: Achievements in space robotics. IEEE Robot. Autom. Mag. **16**(4), 20–28 (2009)
5. Flores-Abad, A., Ma, O., Pham, K., Ulrich, S.: A review of space robotics technologies for on-orbit servicing. Prog. Aerosp. Sci. **68**, 1–26 (2014)
6. Gao, Y., Chien, S., Review on space robotics: Toward top-level science through space exploration. Sci. Robot. **2**(7) (2017). eaan5074
7. Yan, X.T., Tian, X., Brinkmann, W., Palazzetti, R., Melville, C., Li, Y., Bartsch, S., Kirchner, F.: Integrated mechanical, thermal, data, and power transfer interfaces for future space robotics. Front. Robot. AI 5, 64 (2018)
8. NASA.: Mars Exploration Rovers. https://mars.nasa.gov/mer/mission/overview/
9. NASA.: Mars Curiosity Rover. https://mars.nasa.gov/msl/home/
10. Rybus, T.: Obstacle avoidance in space robotics: review of major challenges and proposed solutions. Prog. Aerosp. Sci. **101**, 31–48 (2018)
11. Wong, C., Yang, E., Yan, X.T., Gu, D.: Autonomous robots for harsh environments: a holistic overview of current solutions and ongoing challenges. Syst. Sci. Control. Eng. **6**(1), 213–219 (2018)

12. Wang, Y., Li, D., Hu, C., Wang, Y., Tang, Z., Wang, N.: Review of research on the Chinese space station robots. In: Yu, H., Liu, J., Liu, L., Ju, Z., Liu, Y., Zhou, D. (eds.) ICIRA 2019. LNCS (LNAI), vol. 11743, pp. 423–430. Springer, Cham (2019). https://doi.org/10.1007/978-3-030-27538-9_36
13. Moghaddam, B.M., Chhabra, R.: On the guidance, navigation and control of in-orbit space robotic missions: a survey and prospective vision. Acta Astronaut. **184**, 70–100 (2021)
14. Weber, B., Balachandran, R., Riecke, C., Stulp, F., Stelzer, M.: Teleoperating robots from the international space station: Microgravity effects on performance with force feedback. In: 2019 IEEE/RSJ International Conference on Intelligent Robots and Systems (IROS), pp. 8144–8150. IEEE (2019)
15. Wang, Y.: Current state of space robots. In: Space Robotics, pp. 345–352. Springer, Singapore (2021)
16. Jiang, Z., Cao, X., Huang, X., Li, H., Ceccarelli, M.: Progress and Development trend of space intelligent robot technology. Space: Sci. Technol. (2022)
17. Ding, X., Wang, Y., Wang, Y., Xu, K.: A review of structures, verification, and calibration technologies of space robotic systems for on-orbit servicing. Sci. China Technol. Sci. **64**(3), 462–480 (2020). https://doi.org/10.1007/s11431-020-1737-4
18. Papadopoulos, E., Aghili, F., Ma, O., Lampariello, R.: Robotic manipulation and capture in space: a survey. Front. Robot. AI 228 (2021)
19. Webster, R.J., III., Jones, B.A.: Design and kinematic modeling of constant curvature continuum robots: a review. Int. J. Robot. Res. **29**(13), 1661–1683 (2010)
20. Chikhaoui, M.T., Rosa, B.: Modeling and control strategies for flexible devices. In: Endorobotics, pp. 187–213. Academic Press (2022). https://doi.org/10.1016/B978-0-12-821750-4.00008-6
21. Thuruthel, T.G., Ansari, Y., Falotico, E., Laschi, C.: Control strategies for soft robotic manipulators: a survey. Soft Rob. **5**(2), 149–163 (2018)
22. Della Santina, C., Duriez, C., Rus, D.: Model based control of soft robots: a survey of the state of the art and open challenges (2021). arXiv preprint arXiv:2110.01358
23. Armanini, C., et al.: Soft robots modeling: a literature unwinding (2021). arXiv preprint arXiv:2112.03645
24. Wang, X., Li, Y., Kwok, K.W.: A survey for machine learning-based control of continuum robots. Front. Robot. AI **8**(1), 730330 (2021)
25. Kim, S., Laschi, C., Trimmer, B.: Soft robotics: a bioinspired evolution in robotics. Trends Biotechnol. **31**(5), 287–294 (2013)
26. Wang, H., Totaro, M., Beccai, L.: Toward perceptive soft robots: progress and challenges. Adv. Sci. **5**(9), 1800541 (2018)
27. Stanford press.: Stanford researchers develop a new type of soft growing robot. https://news.stanford.edu/2017/07/19/stanford-researchers-develop-new-type-soft-growing-robot/
28. Mahoney, A.W., Gilbert, H.B., Webster, R.J.: A review of concentric tube robots: modeling, control, design, planning, and sensing. In: The Encyclopedia of Medical Robotics, pp. 181–202. World Scientific (2018)
29. Alfalahi, H., Renda, F., Stefanini, C.: Concentric tube robots for minimally invasive surgery: current applications and future opportunities. IEEE Trans. Med. Robot. Bionics **3202**, 1 (2020)
30. Mitros, Z., Sadati, S.M.H., Henry, R., Da Cruz, L., Bergeles, C.: From theoretical work to clinical translation: progress in concentric tube robots. Annu. Rev. Control., Robot., Auton. Syst. **5** (2021)
31. Shi, C., et al.: Shape sensing techniques for continuum robots in minimally invasive surgery: a survey. IEEE Trans. Biomed. Eng. **64**(8), 1665–1678 (2017)
32. da Veiga, T., et al.: Challenges of continuum robots in clinical context: a review. Progress in Biomedical Engineering, **2**(3), 032003 (2020)

33. Dupont, P., Simaan, N., Choset, H., Rucker, C.: Continuum robots for medical interventions. In: Proceedings of the IEEE. https://doi.org/10.1109/JPROC.2022.3141338
34. Mohammad, A., Russo, M., Fang, Y., Dong, X., Axinte, D., Kell, J.: An efficient follow-the-leader strategy for continuum robot navigation and coiling. IEEE Robot. Autom. Lett. **6**(4), 7493–7500 (2021)
35. Sriratanasak, N., Axinte, D., Dong, X., Mohammad, A., Russo, M., Raimondi, L.: Tasering twin soft robot: a multimodal soft robot capable of passive flight and wall climbing. Advanced Intelligent Systems, p. 2200223 (2022). https://doi.org/10.1002/aisy.202200223
36. Russo, M., Sriratanasak, N., Ba, W., Dong, X., Mohammad, A., Axinte, D.: Cooperative continuum robots: enhancing individual continuum arms by reconfiguring into a parallel manipulator. IEEE Robot. Autom. Lett. **7**(2), 1558–1565 (2022). https://doi.org/10.1109/LRA.2021.3139371
37. Shi, H., Russo, M., de la Torre, J., Dong, X., Mohammad, A., Axinte, D.: Touching the sound: audible features enable haptics for robot control. IEEE Robot. Autom. Mag. (2022)
38. Alatorre, D., Robles-Linares, J.A., Russo, M., Elbanna, M.A., Wild, S., Dong, X., Mohammad, A., Kell, J., Norton, A.D, Axinte, D.: COBRA: From industrial to medical surgery with slender continuum robots. IEEE Robot. Autom. Mag. (2022)
39. Dong, X., Wang, M., Mohammad, A., Russo, M., Ba, W., Norton, A., Kell, J., Axinte, D.: Continuum robots collaborate for safe manipulation of high-temperature flame to enable repairs in extreme environments. IEEE/ASME Trans. Mechatron. (2021). https://doi.org/10.1109/TMECH.2021.3138222
40. Bishop, C., Russo, M., Dong, X., Axinte, D.: A novel underactuated continuum robot with shape-memory alloy clutches. IEEE/ASME Trans. Mechatron. (2022). https://doi.org/10.1109/TMECH.2022.3179812
41. Russo, M., Raimondi, L., Dong, X., Axinte, D., Kell, J.: Task-oriented optimal dimensional synthesis of robotic manipulators with limited mobility. Robot Comput. Integr. Manuf. **69**, 102096 (2021)
42. Angrisani, L., Grazioso, S., Gironimo, G.D., Panariello, D., Tedesco, A.: On the use of soft continuum robots for remote measurement tasks in constrained environments: a brief overview of applications. IEEE Int. Symp. Meas. Netw. (M&N) **2019**, 1–5 (2019). https://doi.org/10.1109/IWMN.2019.8805050
43. Zhang, Y., Li, P., Quan, J., Li, L., Zhang, G., Zhou, D.: Progress, challenges, and prospects of soft robotics for space applications. Advanced Intelligent Systems 2200071 (2022)
44. Mehling, J.S., Diftler, M.A., Chu, M., Valvo, M.: A Minimally invasive tendril robot for in-space inspection. In: The First IEEE/RAS-EMBS International Conference on Biomedical Robotics and Biomechatronics, 2006. BioRob 2006, pp. 690–695 (2006). https://doi.org/10.1109/BIOROB.2006.1639170
45. Wooten, M.B., Walker, I.D.: A novel vine-like robot for in-orbit inspection. In: 45th International Conference on Environmental Systems (2015)
46. Tonapi, M.M., Godage, I.S., Vijaykumar, A.M., Walker, I.D.: A novel continuum robotic cable aimed at applications in space. Adv. Robot. **29**(13), 861–875 (2015)
47. Santiago, J.L.C., Walker, I.D., Godage, I.S.: Continuum robots for space applications based on layer-jamming scales with stiffening capability. IEEE Aerosp. Conf. **2015**, 1–13 (2015). https://doi.org/10.1109/AERO.2015.7118897
48. Nahar, D., Yanik, P.M., Walker, I.D.: Robot tendrils: long, thin continuum robots for inspection in space operations. IEEE Aerosp. Conf. **2017**, 1–8 (2017). https://doi.org/10.1109/AERO.2017.7943940
49. Wooten, M., Frazelle, C., Walker, I.D., Kapadia, A., Lee, J.H.: Exploration and inspection with vine-inspired continuum robots. IEEE Int. Conf. Robot. Autom. (ICRA) **2018**, 5526–5533 (2018). https://doi.org/10.1109/ICRA.2018.8461132

50. Frazelle, C., Walker, I., AlAttar, A., Kormushev, P.: Kinematic-Model-free control for space operations with continuum manipulators. IEEE Aerosp. Conf. (50100), 1–11 (2021). https://doi.org/10.1109/AERO50100.2021.9438380
51. Frazelle, C., Walker, I., AlAttar, A., Kormushev, P.: Encoder-less robot control for space operations with continuum robots. IEEE Aerosp. Conf. (2021)
52. Walker, I.D., Leonard, E.K.: Continuum robot cables and hoses for adaptive inspection and pumping. Int. J. Technol. Eng. Stud. **5**(1), 16–22 (2019)
53. Hawkes, E.W., Blumenschein, L.H., Greer, J.D., Okamura, A.M.: A soft robot that navigates its environment through growth. Sci. Robot. **2**(8), eaan3028 (2017)
54. Blumenschein, L.H., Gan, L.T., Fan, J.A., Okamura, A.M., Hawkes, E.W.: A tip-extending soft robot enables reconfigurable and deployable antennas. IEEE Robot. Autom. Lett. **3**(2), 949–956 (2018). https://doi.org/10.1109/LRA.2018.2793303
55. Nye, T., Milam, M.: Advancements of a MicroSat for On-Orbit Satellite Surgery, Small Satellite Conference 2020. Utah State University, Logan, UT, USA (2020)
56. Dalton, N.M.: Dynamic maneuvers for satellite on-orbit servicing utilizing novel continuum robotics: development and experimentation. Ph.D. dissertation, Western Michigan University (2022)
57. Boschetti, G., Carbone, G., Passarini, C.: Cable failure operation strategy for a rehabilitation cable-driven robot. Robotics **8**(1), 17 (2019)
58. Rodriguez-Leon, J.F., Chaparro-Rico, B.D.M., Russo, M., Cafolla, D.: An auto-tuning cable-driven device for home rehabilitation. J. Healthc. Eng. **1**(2021), 6680762 (2021). https://doi.org/10.1155/2021/6680762
59. NASA. Space Radiation Effects on Electronic Components in Low-Earth Orbit. https://llis.nasa.gov/lesson/824

Design and Performance of a Berthing Space Manipulator

Alexander Titov(✉), Matteo Russo, and Marco Ceccarelli

LARM2 Laboratory of Robot Mechatronics, University of Rome Tor Vergata, Rome, Italy
aleksandr.titov@students.uniroma2.eu, matteo.russo@uniroma2.it, marco.ceccarelli@uniroma2.eu

Abstract. This paper presents a 5R1T manipulator design for berthing tasks with the satellite type CubeSat. The proposed lightweight low-cost solution is capable of holding and transferring objects up to 24 kg. The berthing task is presented to grasp the satellite and moving it to the berthing port in a dynamic simulation using the Autodesk Inventor package. For defined positions of the satellite and the berthing port, the positions of the end-effector are assumed, and the angular positions of the other joints are measured. The problems of grasping and slipping of the satellite are outlined and solved by changing the geometry of the fingertips.

Keywords: Space robotics · On-orbit service robotics · Space Berthing · CubeSats · Space manipulators

1 Introduction

Space exploration starts from Earth orbit. From the first human-made satellite in 1957 to modern times, people made and launched thousands of artificial objects into space [1]. Satellites are nowadays used in many fields, such as telecommunications, navigation, meteorology, observation of the Earth's surface and deep space, and scientific experiments [2]. For satellite operation, low Earth orbit has many advantages over other orbits, such as requiring less energy to launch the object. Therefore, it became occupied by most satellites. However, its volume is smaller when compared with other orbits. This, linked to the fact that failure turns a satellite into debris, means that low orbit capacity is becoming saturated. Currently, there are around 29,000 objects larger than 10 cm on Earth orbit, and most of them are in the low orbits [1]. Collisions between objects in space can destroy them and break them into pieces, producing new debris, which can collide with other objects and cause chain reactions, as described by Kessler [3].

To avoid creating more debris, many solutions have been proposed, such as laser brooms, tethers, nets, and electromagnetic stoppers [4]. These ways are tested experimentally on prototypes, but none of them are widely used. Currently, satellite missions are developed in such a way as to send the satellite back into the atmosphere or into a graveyard orbit by jet engine maneuvers at the end of a mission. Another potential solution is to extend the lifetime of a satellite by refueling or refurbishing it.

© The Author(s), under exclusive license to Springer Nature Switzerland AG 2023
M. Ceccarelli et al. (Eds.): TORVEASTRO 2023, MMS 130, pp. 140–147, 2023.
https://doi.org/10.1007/978-3-031-28447-2_11

Over time, the development of new materials and electronics allowed to build smaller and more cost-efficient satellites. The new standard for the microsatellites that are called CubeSat opened an opportunity not only for large space agencies such as NASA, ESA, or JAXA, but also for small start-up groups and universities to conduct space research [5]. However, the majority of CubeSats are designed for low space orbit, further saturating the limited volume left [6].

For each satellite, lifetime is a key requirement. Fuel consumption to correct orbit, the lifetime of the components, and resistance to space conditions are the main factors that limit the overall lifetime of the satellite [6]. Compared to bigger satellites, CubeSats' lifetime is usually shorter, and they are designed to burn in the atmosphere at the end of the mission [6]. It is, however, possible to extend their life by moving them up and/or refueling them, when necessary, with external systems. To engage and interface with such a small satellite, a manipulator-based strategy, called berthing, can be used.

Current solutions for satellite berthing at the International Space Station (ISS) are Canadarm2, European Robotic Arm (ERA), and Japan Experimental Module Remote Manipulator System [7]. With a body mass greater than 1300 kg, Canadarm2 is capable of handling up to 116,000 kg [7]. The maximum mass of a CubeSat 12U is no more than 24 kg [6]. As such, current manipulators are significantly overengineered for this task. For this reason, this paper proposes a lightweight berthing manipulator for CubeSats.

2 Requirements

A berthing space manipulator is a part of the described berthing system that includes the station, the manipulator itself, and an end-effector [8]. In previous work [8, 9], the requirements of the entire system and for a geometry-based gripper are presented. A berthing task includes reaching the target CubeSat satellite with the manipulator, grasping it with the end-effector, and moving it to the berthing port on the station surface. The accuracy of an end-effector position is influenced by accuracy of all the joints of the manipulator. Referring to existing solutions [7], we obtain 40 mm linear and 1 ° angular accuracies for the 11.3 m ERA manipulator. To grasp a target, the linear accuracy of a developed berthing manipulator is 10% of a satellite's smallest dimension, or 10 mm; angular accuracy is assumed to be the same as for ERA, or 1 °. The manipulator is placed at the station, and it is folded during the launch. For the station in [8], the total length of the manipulator can be assumed as 1.5 m. Its length can help to reach the mentioned precision. However, the manipulator's configuration should allow to reach all the ports with the grasped satellite placed orthogonally to the port. Feedback in joints is necessary to define the relative position of an end-effector with a grasped satellite. For a lightweight solution, the manipulator's mass can be defined less than 5 kg. However, the maximal handled mass is limited by the mass of a biggest satellite CubeSat 12U, or 24 kg plus the end-effector mass of 1 kg.

A summary of requirements for the manipulator can be presented as follows:

- The total length is 1.5 m.
- Ability to reach all the berthing ports in the station.
- The manipulator's total weight is less than 5 kg.

- The handled mass is up to 25 kg [6], where 1 kg is the mass of the end-effector.
- The linear precision is 10 mm, the angular precision is 1° [7].

3 Conceptual Solution

According to these requirements, a kinematic scheme is proposed in Fig. 1a. The manipulator is presented with five rotational and one prismatic joint which can rotate around the station. An extra rotational degree of freedom (d.o.f.) is used for the end-effector. The kinematic scheme is considered with dimensions for CAD design. The dimensions of a conceptual scheme are listed as Denavit-Hartenberg parameters in Table 1. No limitations of the rotation angles θ are set for joints A–E.

The end-effector is presented as a double slider-crank mechanism. A kinematic scheme of half of this mechanism is shown in Fig. 1b To grasp the satellite, L-shaped fingertips of the end-effector are connected to the slider. The angle between surfaces of the fingertip is 90°. When the end-effector is in open position, the distance between fingertips is larger than the biggest satellite's diagonal. Rotating the motor to 90 degrees moves the fingertips to grasp the smallest satellite. To carry out the grasping for the CubeSats from 1U to 12U by its ribs, the distance between fingertips should vary in the following way. The largest distance is assumed to be 5% larger than the diagonal of the largest satellite, 320 mm. So, the distance from the centre to the fingertip GK_1 is equal $d/2 = 168$ mm. The end-effector is defined to be closed when the central crank rotates $\alpha = 90°$. In this case, the crank GHs that is connected to the finger is a diagonal of a triangle GHK and a minimal distance $l_0 + h$ equal to 70 mm, h is assumed 25 mm. The length of the central crank l_1 is 64 mm, the other crank l_2 is 78.6 mm long.

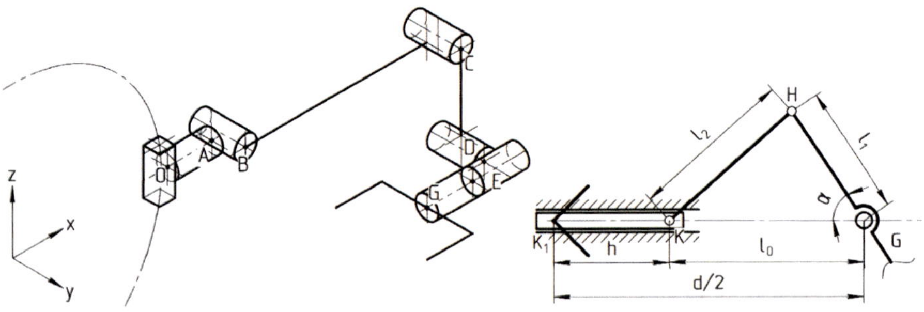

(a) (b)

Fig. 1. Kinematic schemes of: (a) manipulator, (b) end-effector

Table 1. Denavit-Hartenberg parameters for the kinematic scheme in Fig. 1

	O	A	B	C	D	E	G
a (m)	0.8	0.1	0.04	0.9	0.7	0.04	0.03
α (°)	0	90	90	0	0	−90	0
d (m)	0	0	0.03	0.015	0.045	0.03	0
θ_{ref} (°)	0	−90	−90	−90	0	90	0

4 A CAD Design

Following the kinematic scheme from Fig. 1, a model is designed in CAD Autodesk Inventor. The manipulator is installed on the station and can move around it. The translational motion is provided by the wheel which is in contact with a rail of the station. The diameter of a wheel is 50 mm, and the rail diameter is 730 mm. The rotational joints are encased in 60 × 60× 30 mm units with two connection holes: one is for rigid connection; another is for rotation. The reference frame is set in the center of the rotation hole, Z is the rotation axis. Two links have dimensions 890 and 610 mm with a motor fixed on one top. Fingertips' dimensions are set as 100 × 30 mm. For a better contact with the satellite, the fingertips are designed with 16 half-spherical bulges on each surface.

Aluminum is chosen as the material for the manipulator's parts. The entire mass of the manipulator is 3.98 kg. The CAD design is presented in Fig. 2.

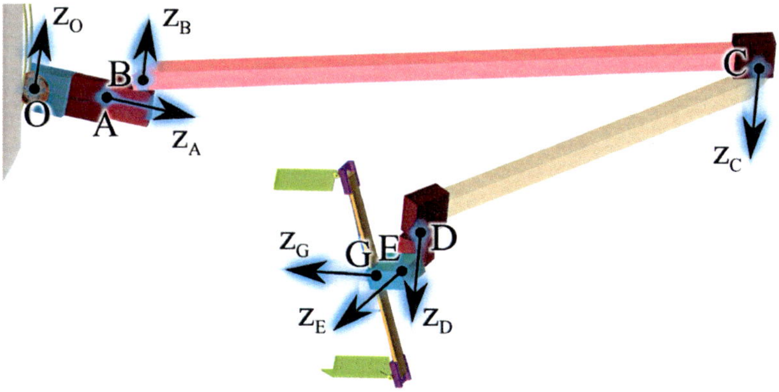

Fig. 2. A conceptual solution of a manipulator design from Fig. 1

5 Performance Analysis via Simulation

The model in Fig. 2 has been used to simulate a berthing task with Autodesk Inventor. The selected task starts with identifying the position of the satellite. From the folded configuration, the manipulator reaches for the satellite, grasps it with the end-effector, moves it to the berthing port and, finally, places it there. This operation is characterized by six path points (P0–P5):

1. P0. The manipulator and the end-effector are in a folded configuration.
2. P1. Aiming position: end-effector is 50 mm beneath the satellite, end-effector is opened, and fingertips are parallel to the ribs of the satellite.
3. P2. Ready to grasp. Fingertips' edges are 30 mm above the satellite's bottom.
4. P3. Grasping the satellite by its ribs with the end-effector's fingertips.
5. P4. Transferring the satellite to the berthing port. The satellite is centered, its surface is parallel to the port, the distance between satellite and port surfaces is 50 mm.
6. P5. Insert the satellite into the berthing port. The surface of the satellite and the berthing port are aligned.

By moving the model's end-effector to each of those points, the respective joint angles φ have been numerically evaluated, as presented in Table 2. Angle φ_O for a prismatic joint is calculated from folding position of the manipulator. For every position, the operation end time t_{end} is presented in the last column of Table 2. The snapshot of the berthing process is shown in Fig. 3.

Table 2. Input joint angles φ for the berthing task

	φ_O (°)	φ_A (°)	φ_B (°)	φ_C (°)	φ_D (°)	φ_E (°)	t_{end}(s)
P0	0	−180	90	180	−180	90	0
P1	175.87	0	51.2	146.61	5.41	94.13	19
P2	175.87	0	51.33	139.39	−1.9	93.44	20
P3	175.87	0.63	51.33	139.39	−1.9	93.44	22
P4	290.41	32.02	−64.6	99.52	−138.54	130.79	35
P5	291.57	31.38	−66.03	99.55	−137.7	131.97	36.5

The position of the prismatic joint O travels along a circular rail controlled by a rotational motor; motion can be transmitted with a gear coupling, friction gears, or an equivalent transmission system. By defining the transmission ratio of this driving system, φ_O can be computed. With the provided example geometry, a rotation of $175.87 \div 50 \times 730 = 2657.7°$ moves point O by a 175.87° angle. With a start and end point of zero and an acceleration of 360°/s^2, it is assumed that the maximum angular velocity is 360°/s before grasping. During manipulation, this speed is limited to 10°/s to avoid problems due to the larger inertia.

According to Table 2, the grasping operation itself is realized in steps P2 and P3, and it lasts 2 s. The motion law for input joint G of an end-effector is presented in Fig. 4. The central crank initial position set up to 3° to avoid dead points. End position is 89.5° to grasp the satellite 1U. To decrease shocks, movement from 80 to 89.5° is slower than from 3 to 80°. At this point, the imposed motion is turned off when one of the contact forces between the fingertip and the satellite exceeds 20 N, and the joint starts working

Fig. 3. A snapshot of a berthing task: folding, grasping, insertion

as a torsion spring. The parameters of a spring are defined in free position of 90°, elastic stiffness of 100 N·mm/°, and a coefficient of dry friction in the joint G equal to 0.3. The results are presented for the input torque in joint G in Fig. 5, whereas the simulated contact forces between the satellite and fingertips are reported in Fig. 6.

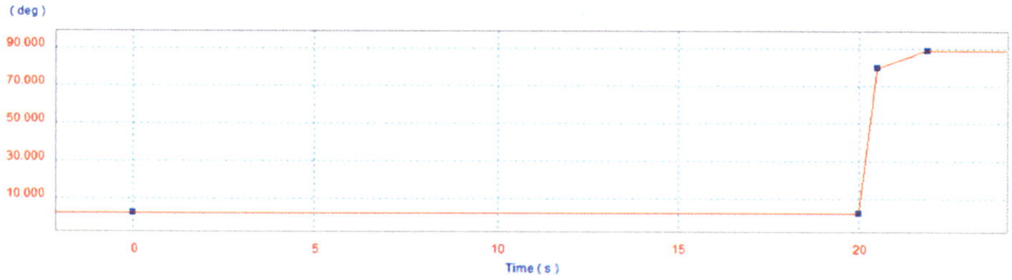

Fig. 4. A screenshot of a motion law definition for input joint G from Fig. 1

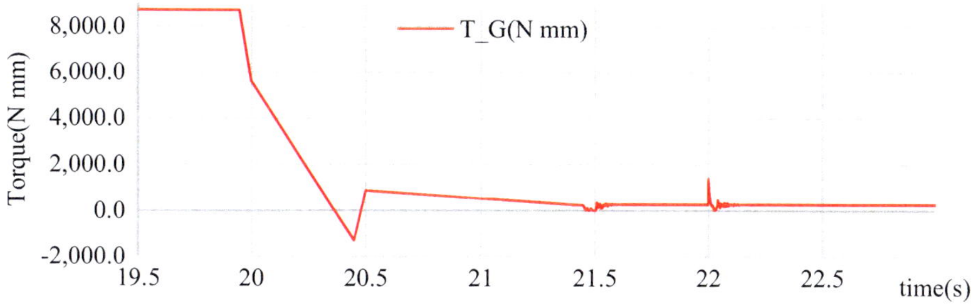

Fig. 5. Computed results of an input torque during grasping

For Fig. 4, the imposed motion and a spring work together, and to keep the position of a crank at 3°, the motor needs 8700 N·mm torque. When grasping, crank goes faster than the spring returns, and torque of 1760 N·mm acts to the other direction. At the end of the grasping operation, the imposed motion is turned off and the satellite is fixed with a torque of 255 N mm.

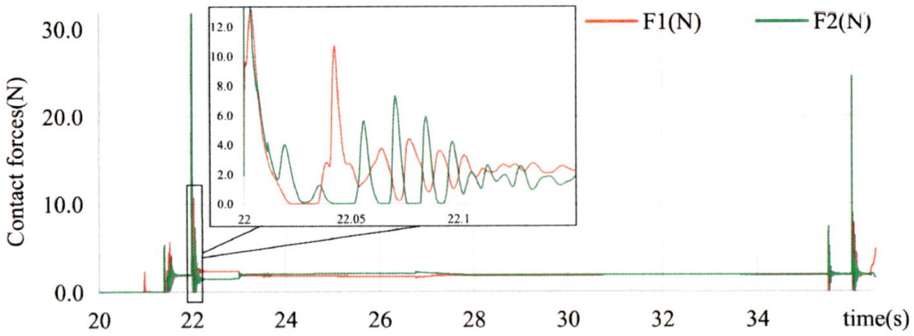

Fig. 6. Computed results of contact forces in fingertips during grasping and moving to the berthing port

In Fig. 6, contact forces are shown for each fingertip. The contact does not occur at the same time for both fingertips, so the peak of 2.5 N for one fingertip is presented at 21 s. The highest peak contact forces for both fingertips occur at the time when the satellite is aligned with the profile of the fingertips. From 22 s, manipulator starts to move the satellite. Accelerations in joints are not smoothened for this test; therefore, the highest contact forces occur at this time. During the movement of the manipulator, the satellite is aligned with it, and the contact forces of both fingertips converge. Two other peaks occur at the end of the berthing process, when the satellite contacts the berthing port.

In Fig. 7, the displacements of the point G in the center of the end-effector, are presented as referenced to the center of the bottom surface of the satellite, with the x-axis directed along one of the diagonals of its square surface. This reference frame shows the aiming of the end-effector before and during the grasping.

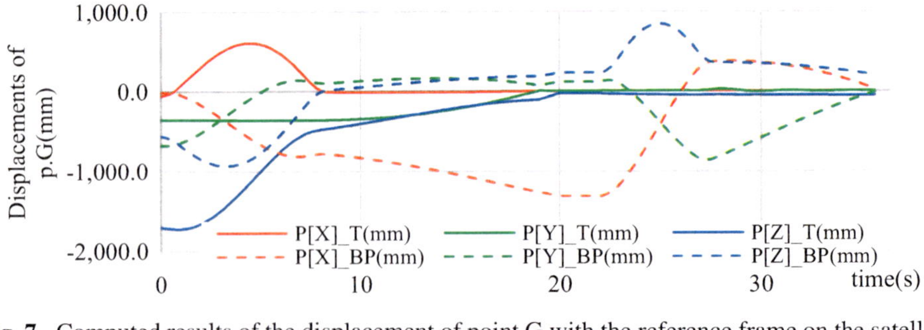

Fig. 7. Computed results of the displacement of point G with the reference frame on the satellite (T) and the berthing port (BP)

After grasping, the satellite is held by the fingertips, and coordinates P[X]_T and P[Y]_T are near zero. As such, this operation is referenced to a frame set in the middle of the berthing port of the station. Therefore, the operation is successful once both X- and Y-coordinates are near zero, with the Z-coordinate here representing the distance between those two local frames.

6 Conclusions

A berthing space manipulator is designed for CubeSat microsatellites. The performance of the design is evaluated with a dynamic simulation in Autodesk Inventor. Inverse kinematics is solved for six positions, from the folded configuration of the manipulator up to the insertion of the satellite into the berthing port. The method presented here works using a known satellite position. The problems of grasping and holding the satellite are indicated. Flat surfaces of fingertips do not fix the satellite, and it slides off when moving to the berthing port. The addition of a limited number of contact points on the fingertip's surface is necessary to ensure the contact and prevent slipping. Performance analysis is carried out for CubeSat 1U. The described method for berthing can be applied to satellites of other types as well.

References

1. ESA Space Debris Office. ESA's annual space environment report. ESA ESOC, Darmstadt, Germany (2022)
2. Kodheli, O., et al.: Satellite communications in the new space era: a survey and future challenges. IEEE Commun. Surv. Tutor. **23**(1), 70–109 (2021). https://doi.org/10.1109/COMST.2020.3028247
3. Kessler, D.J., Cour-Palais, B.G.: Collision frequency of artificial satellites: the creation of a debris belt. J. Geophys. Res. **83**(A6), 2637–2646 (1978)
4. Mark, C.P., Kamath, S.: Review of active space debris removal methods. Space Policy **47**, 194–206 (2019). https://doi.org/10.1016/j.spacepol.2018.12.005
5. Lu, S., Theoharis, P. I., Raad, R., Tubbal, F., Theoharis, A., Iranmanesh, S., Abulgasem, S., Khan, M. U. A., Matekovits, L.: A Survey on cubesat missions and their antenna designs. Electronics **11**(13), 2021 (2022). https://doi.org/10.3390/ELECTRONICS11132021
6. Johnstone, A.: CubeSat design specification Rev. 14.1 the CubeSat program. Cal Poly SLO pp. 1–34 (2022)
7. Laryssa, P., Lindsay, E., Layi, O., Marius, O., Nara, K., Aris, L., Ed, T.: International space station robotics: a comparative study of ERA, JEMRMS and MSS. 7th ESA workshop on advanced space technologies for robotics and automation "ASTRA 2002". Noordwijk, The Netherlands, pp. 1–8
8. Titov, A., Ceccarelli, M.: Requirements and problems for space berthing system. In: Advances in Italian Mechanism Science—Proceedings of the Fourth Conference of IFToMM ITALY, paper 005 (in print) (2022)
9. Titov, A., Ceccarelli, M.: Design and performance characterization of a gripper end-effector for a space berthing manipulator. In: Mechanism and Machine Science—Proceedings of MeTrApp2023, paper 007 (submitted) (2022)

Check for updates

How Swarm Robot Dynamic Can Describe Mechanical Systems

Ramiro dell'Erba(✉)

ENEA Robotics and Artificial Intelligence Laboratory, 00123 Roma, Italy
ramiro.dellerba@enea.it

Abstract. In this paper we shall see how a study, started from flocking rules normally used to spatially coordinate a large number of cooperating robots, can be used to solve deformation problems of a continuous material. The advantage is to save compute time with respect of traditional solution of Newton's equations. To this aim we shall impose kinematic rules between the material points as if they were component robots of a swarm and, as such, with precise distancing behaviours borrowed from bird starlings. The approach used is purely kinematic and makes no claim to accuracy but provides very plausible results. Although the force of gravity is not often present in the space domain the deformation of materials, due to the large temperature range, is a hot topic for robotic systems used in that domain. The model is able to reproduce the behaviour of deformable bodies with standard or generalized (Cauchy or second gradient) deformation energy density. The tool gives a plausible simulation of continuum deformation also in fracture case. Some of the material has already been published, but there is an important change from the old papers: a new calculation technique has been used that allows significant progress and appears to be very promising.

Keywords: Swarm robotics · Mechanic · Deformation

1 Introduction

1.1 The Origin of Idea

Lessons from Nature can be very useful in designing the control system of a robotic swarm; in our laboratory, we have been working on this issue for many years. In [1, 2] the authors studied how to obtain the geometric configuration of a submarine robotic swarm by knowing the distance between elements; this is a very important task because the swarm, like fish in a school, adapts its configuration depending on the assigned mission. The concept of robotic swarming has been a hot topic in the robotics scientific community for many years. Research has been inspired by the biological behaviours of animals, such as bees [3, 4], exploiting the concepts of social activity [5], division of labour, information sharing, and cooperation in the tasks to be performed. The advantages of using a swarm are many. A single robot is prone to failures that can jeopardize the success of the entire mission. A multi-robot approach can benefit from parallelism of operations and

© The Author(s), under exclusive license to Springer Nature Switzerland AG 2023
M. Ceccarelli et al. (Eds.): TORVEASTRO 2023, MMS 130, pp. 148–159, 2023.
https://doi.org/10.1007/978-3-031-28447-2_12

redundancy. Also, what is not possible at a single robot, the operator is offered multiple views simultaneously, putting him in a position to use gradient techniques to track, for example, oil leaks. In a swarm, members operate with a common goal, sharing the workload; the absence of one member can be easily handled by redistributing the work among other members. A swarm can be seen as a single body, interfacing with human end users as a single creature and overcoming the problem of controlling a large number of individuals. There is no central brain, but each individual can possess an intelligent local control system that can manage its own choices by observing what its neighbours do. For this purpose, a positioning and control algorithm was developed to achieve the desired configuration. We then used the same algorithm to describe the behaviour of complex objects subjected to deformation.

1.2 State of Art

A system of material particles evolves over time following Newton's laws of dynamics; however, in recent times, with the evolution of Computer Graphics applied to video games, a great interest has developed in the study of a system of particles whose motion is determined by their relative position, without solving the differential equations of dynamics. The name of this method is position-based dynamics (PBD) [6]. This method considers internal forces in the system as constraints on positions, and thus the mutual interaction between the particles is explained as relations between positions. External forces are left as such. The goal is to simulate the behaviour of real materials to obtain graphically realistic results, to the user's eye. Physical principles are partially applied through ad hoc heuristic assumptions and are often implemented without direct computation of equations to save computational time. PBD methods produce physically plausible behaviour, but suffer from limitations in modelling complex material properties and describing interactions between heterogeneous bodies. The goal of PBD is not to model physical processes exactly, but to generate visually plausible simulation results with low computational cost and sacrificing some accuracy compared to solving equations with classical finite element methods (FEM).

Knowledge of traditional forces is then abandoned in favour of knowledge of positional displacements; the problem is then transformed into a geometric constraint between configurations. The computation can be parallelized on GPU cores for greater efficiency.

Basically, in PBD, the displacement of a particle (discretized continuum) is determined by the position of its neighbours. Therefore, the computation of the new position of a set of particles can be viewed as a constrained geometric problem leading to a transformation operator between the matrices representing the particle configuration, C_t, for a discrete set of time steps $t_1, t_2, \ldots t_n$. It should be noted that most PBD methods hide the dynamics within their relation; they also require knowledge of the particle velocity, which we try to avoid.

We have written a complete, customizable, modular algorithm that is easily expandable for whatever new features we want to introduce. One of the main advantages of the proposed algorithm is that it automatically accounts for elasticity under large deformations, a topic that plays an increasingly important role in today's research.

The goal of the proposed model has been to develop a numerical tool that can model the behaviour of deformable bodies and able to take in account for higher gradient constitutive relationships. The main difference between our proposed method and the PBD, lies in the use of particle velocity, often used in the PBD; this, in our opinion, hides the internal dynamics, therefore we have avoided to use it. Since our algorithm is based on a linear operation, its computational cost increases only linearly with the number of particles in the system. In [7, 8] it has been shown how the model can exhibit a rich range of behaviours, such as asymptotic convergence to the equilibrium configuration, instabilities of various types and, under well-defined circumstances, spontaneous growth of the crack length after a quasi-stationary state. So far, we have tried to describe the deformation of a continuous medium with this useful tool for complex microstructures not easily analysed by Cauchy continuum theory [9, 10]. Classical Cauchy continuums cannot provide accurate predictions in highly inhomogeneous microstructures [11, 12], although some generalizations have been introduced. The technological interest in the development of exotic mechanical metamaterials capable of performing targeted tasks make the study of new and efficient algorithms of great interest. These materials can be produced by simple three-dimensional (3D) printing. The amount of new experimental data opens up numerous complex theoretical problems; so far, it is clear that, in this context, numerical tools are essential to have an adequate mediation between theoretical and experimental results.

2 The Algorithm

In this paper we limit ourselves to a two-dimensional continuum. The continuum can be divided into a finite number of particles that occupy, in their initial configuration, the nodes of a lattice. The type of lattice is chosen from the five Bravais lattices; changing the lattice (and/or other parameters) can produce different results under the same initial conditions. In summary, leaders move following an assigned trajectory. Immediately thereafter, followers follow them according to a rule (e.g., positioning themselves in the center of gravity determined by their first neighbors). To avoid edge effects, we create a frame of outer points, with its own rule of motion; typically, it translates the follower of the nearest point. Finally, to handle fractures, we insert ghost points if the distance between one point and others exceeds a certain threshold. For a full description of the algorithm, see [13]. Typically, the number of first neighbors is *nc*, where *nc* is the lattice coordination number. As an example of a rule, one may decide to use the barycenter; in this case, the new x-coordinate of particle *j* is

$$x_j(t) = \frac{\sum_{k=1}^{all\ neighbours\ of\ j} x_k(t)}{N} \tag{1}$$

where N is the total number of neighbors, *nc*; a similar equation can be used for the *y*-coordinate. In this way, the displacement of a follower point is the average value of the displacements of its first neighbors; the number of shells determines the order of interaction. We can use different rules to mimic different constitutive equations. Possible generalizations of Eq. 1 include the use of the geometric mean, the mean raised to power, and the weighted mean. One of the possible weights is the Euclidean distance *dis(k, j)* between particle *k* and its neighbor *j*. This can simulate Hook's law, in which the pullback force increases with increasing strain. According to Eq. 1, since the *x* and *y* coordinates are independent, the Poisson effect cannot be obtained. One possibility to get it is to use

$$y_j(t) = K * (x_j(t) - x_j(t_0)) * da + \frac{\sum_{k=1}^{all\ neighbours\ of\ j} y_k(t)}{N} \tag{2}$$

where *da* is a function of the distance from the central axis, *K* a parameter determining the response force and $x(t_0)$ the *x* coordinate at time t_0. So far modification of *x* coordinate has effect on the *y* coordinate. The Euclidean distance, *dis(k, j)*, can be used as weight.

$$x_j(t) = \frac{\sum_{k=1}^{all\ neighbours\ of\ j} dis(k,j)x_k(t)}{\sum_{k=1}^{all\ neighbours\ of\ j} dis(k,j)} \tag{3}$$

We can also force the motion of the followers to exceed the equilibrium position of the centre of gravity by forcing the particles to oscillate.

$$x_j(t) = \frac{\sum_{k=1}^{all\ neighbours\ of\ j} w(k,j)x_k(t)}{\sum_{k=1}^{all\ neighbours\ of\ j} w(k,j)} + fd\left(\frac{\sum_{k=1}^{all\ neighbours\ of\ j} w(k,j)x_k(t)}{\sum_{k=1}^{all\ neighbours\ of\ j} w(k,j)} - MT(j, t_0)\right) \tag{4}$$

where *w(k, j)* is the weight, *fd* is a feedback factor and $MT(i, t_0)$ is the *x*-coordinate of *j* point at t_0 to have memory of the initial configuration.

When the leaders move, the followers are no longer in the equilibrium position, given by Eq. 1. We can calculate, for each pseudo time step, the new lattice configuration by four operations in sequence. Leader movement, follower movement, frame movement and, finally, fracture management. Thus, we can say that leader motion "propagates" through the lattice to affect the position of all followers in a "pseudo-time" interval that depends on the size of the lattice and the number of shells considered in defining neighbours. Similarly, when the leaders stop, the followers continue to adjust their position in many pseudo-time steps. To avoid edge effects, we construct a frame surrounding the body with an outer shell of points, so that each follower interacts with the same number of elements. Without the frame, a corner point has fewer neighbours than an interior point and (according to Eq. 1) would collapse inward. The motion of the frame is simple: it only has to follow the motion of one of its assigned followers; in cases where there are more assigned followers (e.g., in a corner), an average displacement, or a more general complex rule, can be applied.

The model exhibits pronounced nonlinear behaviours, as shown in [14], because the composition of motions for the leader does not lead to a simple superposition of effects in the system configuration.

To handle fracture phenomena, it is assumed that interactions decrease with increasing distance between particles. Therefore, when the Euclidean distance between points is "large," say greater than a fixed threshold, they lose their interaction. These "distant" elements stop influencing each other and are no longer considered in the calculation of the follower's position. For the purpose of balancing the point displacement calculations we introduce ghost points, otherwise we have the usual problem of an incomplete crown of neighbours. The typical position where ghost points are placed is the one that allows us to recover the original shape of the lattice however, other possible choices lead to different results. All the properties of these ghost elements are the same as those of the followers, but their movement is not considered because they are not in the list of followers. They are just in the right position to balance the cell. As we have seen, a change in their settling position produces effects such as contraction or loosening of the lattice in the deformed configuration. In fact, by varying the distances of the ghost elements after fracture from the true elements, plastic and elastic behaviours can be obtained. Elastic behaviour in fracture refers to the property of fracture edges or disconnected pieces originated after fracture to recover their original shape.

The computation of the new position for a set of particles can be viewed as a constrained geometric problem using a transformation operator between the matrices describing the configuration of the particles, C_t, for a discrete set of time steps t_1, $t_2, \ldots t_n \ldots\ldots$

In practice, it is necessary to decide the lattice constraints and interaction rules between the followers, to imitate the correct behaviour of the constitutive equations of materials.

In the near future we are considering the possibility of discussing the proposed model in a fully variational setting, which is by no means trivial but would provide clear methodological advantages.

Moreover, an algorithm based on the geometric centroid of the neighbours of a given particle is consistent with the idea of locally minimizing an elastic potential, since the centroid has the well-known property of minimizing the sum of squared distances from a set of data points in a Euclidean space. Therefore, the proposed algorithm seems a natural discrete approach from the variational point of view. Another possibility, close to these concepts, that we are considering is to replace Eqs. 1–4 with a potential field that can determine particle displacements and avoid collisions.

Another important possibility, which is the main novelty presented in this paper, is to use an optimization method that minimizes the square of the distance between all particles in the lattice simultaneously. The difference of this method, compared with the previous one, is fundamental. In fact, in this way it is possible to direct the optimization by imposing constraints between the coordinates of the particles themselves or some of them (e.g., fixed distances for some of them), in order to consider complex structures such as pantographs, sets of beams or the like. The optimization must then search for the best coordinates consistent with the imposed conditions, which lend themselves to describing a wide variety of situations. For example, I can search for coordinates with the constraint that the distance between two particles is fixed, signifying that these two are part of a fixed axis. The objective function, the sum of the squares of the distances, is not random. It tends to mimic the potential energy of an elastic system. Other objective functions with other meanings are possible.

3 Some Examples

Let us now address some two-dimensional problems concerning objects of simple shape whose leaders are subject to imposed displacements of significant physical meaning; we have chosen two examples able to show the great potentiality of our tool.

Case (1) Simple stress test with fracture.

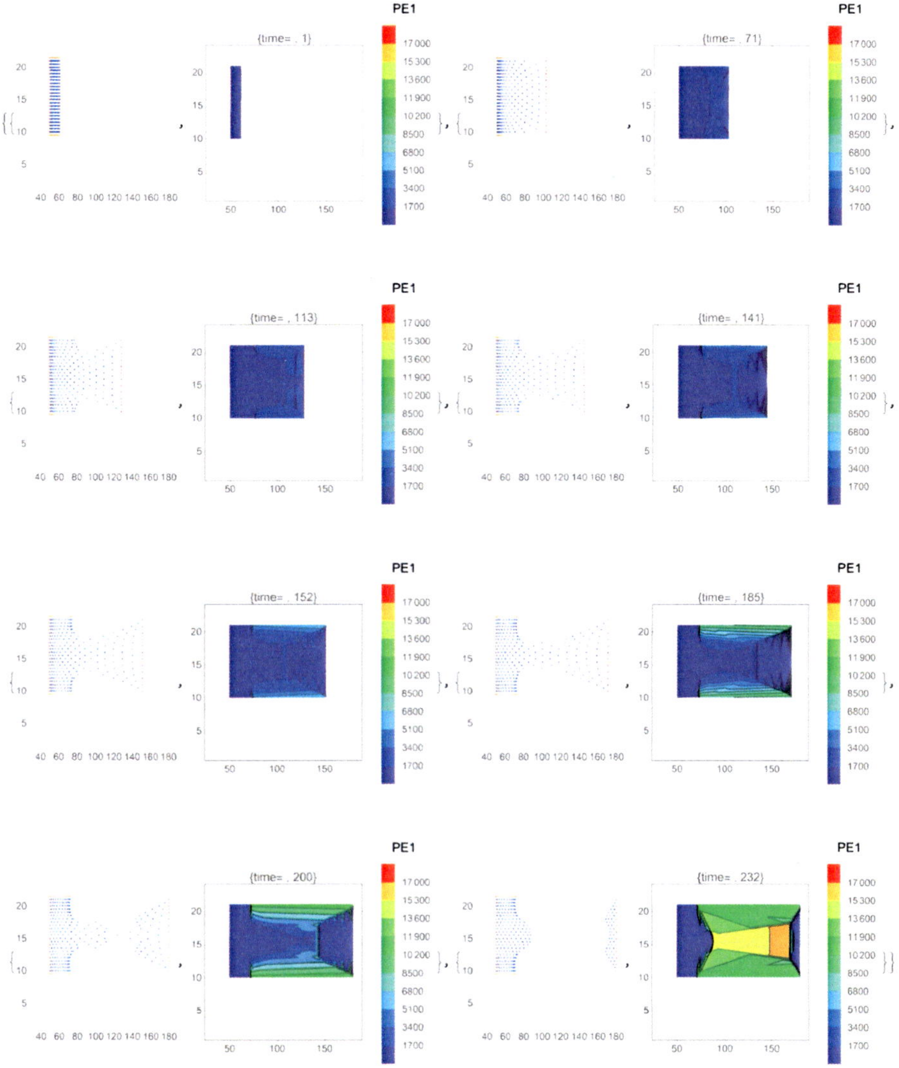

Fig. 1 Tensile test with fracture rectangular centred lattice, coordination number 5. Configuration over different time (1, 23, 33, 38, 40 and 63).

Consider a square-shaped specimen subjected to tension in a tensile test. The tension is conducted from the upper right side, while the other side is blocked, at constant velocity in the x direction (velocity 0.6 length/time units); the lattice chosen is a centred rectangle of 10×10 particles. The number of neighbours, *nc*, is five and a first gradient interaction is considered.

Figure 1 shows the result of the tensile test; the red dots are the leaders, the blue the followers and the orange the frame. The simple rule governing the movement of the followers is that each point must be at the centre of gravity of its neighbours (Eq. 1). The low number of first neighbours chosen makes the deformation much smoother, allowing the detachment of a large number of particles, as can be seen in the right side of the images. The behaviour is therefore more plastic. A modification of the parameters showed how this deformation can be transformed into a brittle deformation.

Note how the non-detached points, having lost contact with the leader, return to their original configuration.

An interesting phenomenon is observed when considering an oblique lattice (Fig. 2), because of the asymmetry (see the five red leaders on the right) of the leaders with respect to the frame. In fact, if we consider an axis of symmetry in the x-direction, we can see two leaders near the frame in the upper level and only one near the lower level. Thus, an interesting fracture behaviour can be observed. The fracture starts from the bottom, where the attraction of the leaders is less, and propagates vertically. It resembles a tear on a sheet of paper (Fig. 3). The distance of the fracture is 10 units and the velocity is 0.4 units per time step. Again, the purpose is to show the great versatility of our instrument and its ability to mimic complex behaviours with simple rules.

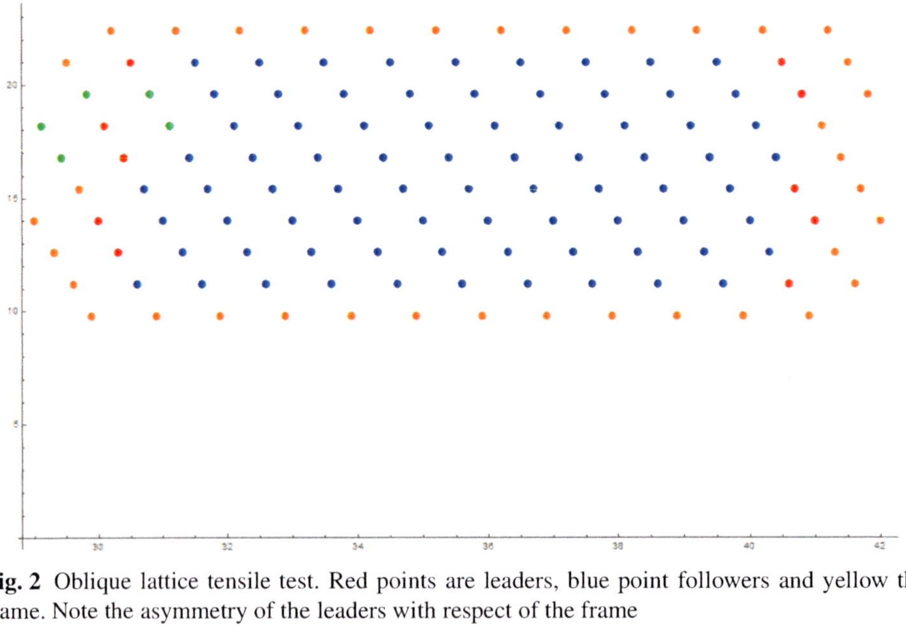

Fig. 2 Oblique lattice tensile test. Red points are leaders, blue point followers and yellow the frame. Note the asymmetry of the leaders with respect of the frame

At this point we make an important variation in the method used. We consider the set of all points and find the Cartesian coordinates that minimize the objective function "distance squared" of the particles to each other. The minimization operation has as its solution all Cartesian coordinates simultaneously. Comparing the results with previous tests shows that it is possible to obtain results similar to those in which the "barycentre rule" (Eq. 1) is applied. In fact, the barycentre enjoys precisely the advantage of minimizing the square of the distance of the particles from each other. However, this method has the advantage of easily lending itself to the introduction of constraints between the

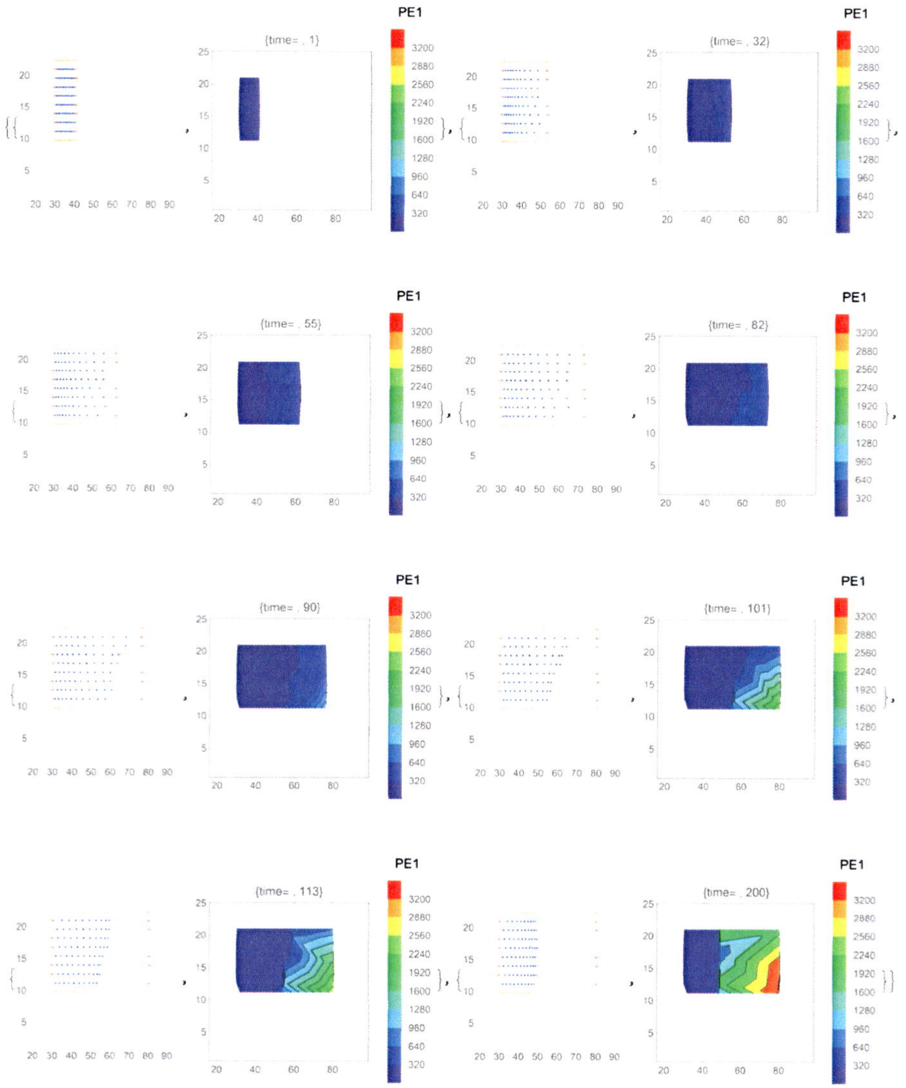

Fig. 3 Tensile test with oblique lattice: breakage fracture. Configuration over different time (1, 32, 55, 82, 90,101,113 and 200).

particles themselves, which is not easy in the previous case where new coordinates are calculated point by point.

We therefore want to examine the behaviour of a pantograph composed of inextensible beams held together by pins. The beams are free to flex. In Fig. 4, the beams are highlighted in yellow in the undeformed configuration. If we now perform a pull from the right side, the final configuration is as shown in Fig. 5. The constraint between the yellow points belonging to the inextensible beams remains. Pantographic structures are very important for generating highly nonlinear behaviour by small locally linear deformations; therefore, a descriptive model of them is useful in the context of structured materials used in the aerospace environment.

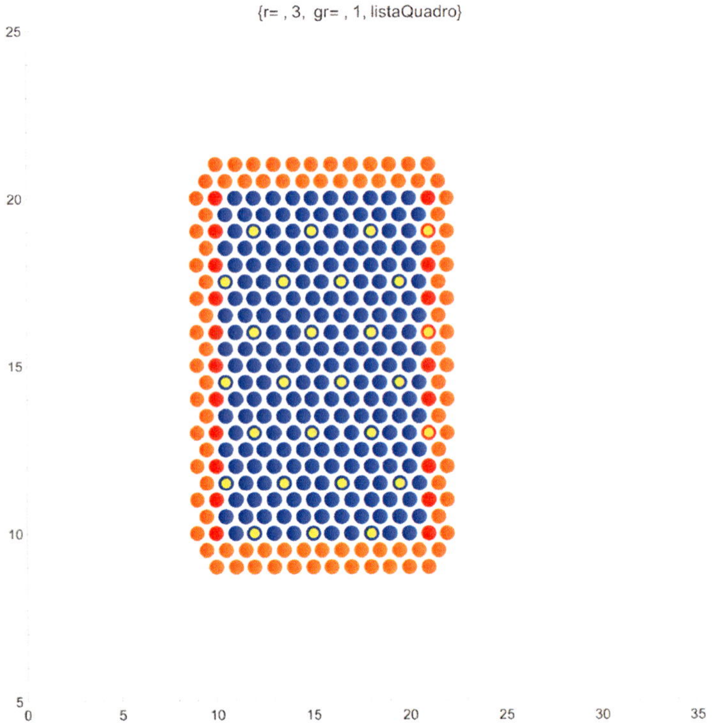

Fig. 4 Oblique lattice tensile test with constrained. Yellow points belong to unextendible beam and are the pivot. Undeformed configuration

4 Conclusion and Future Work

In this work, we have discussed about a tool, presented in previous works but with a new methodology here, able to describe strain deformation of a continuum medium taking in account complex physical effects. Flocking rule, used in robot swarm, can be used to describe deformation of bidimensional continuum medium using a simple algorithm able to show account complex physical effects in a plausible way. The algorithm is highly customizable. This software tool is based on position-based dynamics, but differently from it we do not ask for the knowledge of the velocity and do not introduce forces. The strain is imposed on some particles (leaders, component of the swarm) whose motion is assigned and the other particles (followers) move themselves according to kinematic rules. The motion of the followers, like in a bird swarm, is determined by the position of their neighbours. The tool is based on Position Based Dynamic. Differently from the PBD methods used in computer graphics, we still do not ask for the knowledge of the

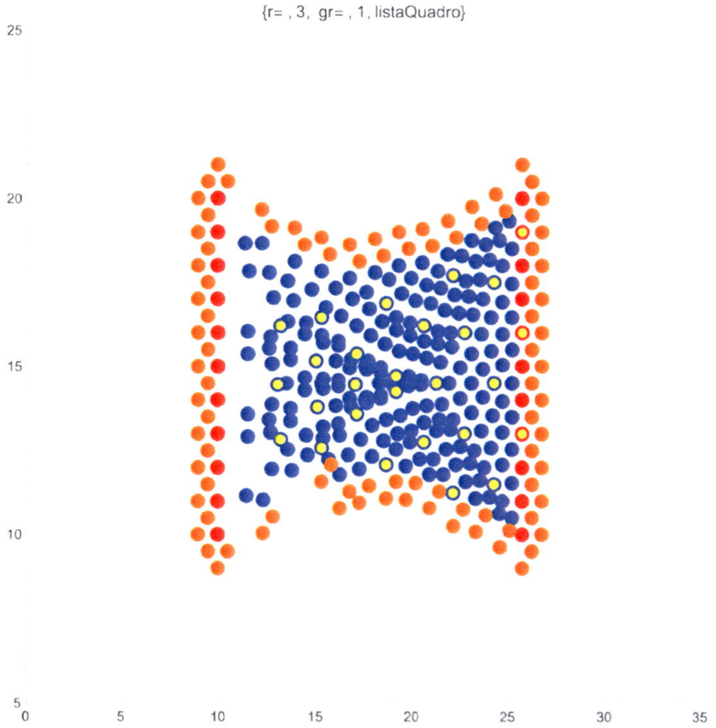

Fig. 5 Oblique lattice tensile test with constrained. Yellow points belong to unextendible beam and are the pivot. Deformed configuration; note the beam design.

velocity and do not introduce some kind of forces to take in account mechanical effects. So far, the deformed configuration is calculated not by Newton's law but only by the relative positions between the particles of the system, the characteristics of the lattice and by rules describing the how a particle would like to place with respect to its neighbours. Changes of some parameters like, lattice interaction rules, fracture distance, numbers of neighbours (i.e., to introduce first and second gradient theory) lead to different behaviour.

In a preceding work we verified the results of this tool are in accordance with results obtained by FEM, also in the fracture case; the results showed good similarity between the prediction of the discrete model and the predictions of standard FEM simulations.

Being based on a linear operation the computational cost of the algorithm increases linearly with the number of the elements; on the contrary, usually, cost usually associated with FEMs has over-exponential growth.

Essentially, we have to compute the action of a transformation operator between matrices and the job can be parallelized between the GPU cores of the powerful video card, saving computational cost need to solve FEMs.

This is not a new kind of physics, just a graphic representation of a plausible behaviour; keep in mind that, up to now, you do not start from the constitutive equations of the materials leading to the rules governing points displacement. Actually, we just imitate a known behaviour adjusting the algorithm parameters. Anyway, the results are also interesting even if still in a preliminary form.

The presented results are interesting but they still are at a preliminary stage. We have collected some success showing plausible deformation in different conditions but when we consider a beam under loading the results are not good and the need to connect constitutive equations with the parameters of our tool emerges powerfully. So far, the most important topic to be investigated is concerning how to relate materials parameters with the choices we are doing in our tool. As matter of fact we, actually, have not criteria about how direct our tool's choices to describe a particular material continuum.

Generalization in 3D of the tool is quite easy but still needs some optimization in the code to keep the computation time in the order of seconds, by using a normal PC Desktop.

We are relaxing the hypothesis that the neighbours always are the same to describe liquid and gas; this needs an intermediate calculation step because you have to compute who are the neighbours, defined in this case as the particles inside a specified volume, at each time step.

The most important innovation, introduced in this work, is the calculation performed not point by point but considering all points together through an optimization procedure. This opens the door to the possibility of considering structured object like pantograph, which are extremely useful in aerospace engineering, so that constraints can be entered much more explicitly and easily. It can be described as a set of beams with constrained point in the pivot or, in the Hencky vision, can be conceived as a set of points interconnected by springs.

Further developments are concerning different fracture mechanism, different frame to avoid edge effects, other interactions rules and adaptive lattice.

An appropriate potential interaction could be useful for a better description of different deformation regimes, such as elastic and plastic ones together with fracture.

References

1. dell'Erba, R.: Determination of spatial configuration of an underwater swarm with minimum data. Int. J. Adv. Rob. Syst. **12**(7), 97 (2015). https://doi.org/10.5772/61035
2. dell'Erba, R.: Distance estimations in unknown sea underwater conditions by power LED for robotics swarms. Continuum Mech. Thermodyn. **33**(1), 97–106 (2020). https://doi.org/10.1007/s00161-020-00889-x
3. Karaboga, D.: An idea based on honey bee swarm for numerical optimization', Technical report-tr06, Erciyes university, engineering faculty, computer engineering department. [Online]. http://www-lia.deis.unibo.it/Courses/SistInt/articoli/bee-colony1.pdf (2005). Last accessed 07 Nov 2014
4. Passino, K.M., Seeley, T.D., Visscher, P.K.: Swarm cognition in honey bees. Behav Ecol Sociobiol **62**(3), 401–414 (2007). https://doi.org/10.1007/s00265-007-0468-1
5. Khatib, O., Kumar, V., Rus, D.: Experimental Robotics: The 10th International Symposium on Experimental Robotics. Springer Verlag (2008)
6. Bender, J., Koschier, D., Charrier, P., Weber, D.: Position-based simulation of continuous materials. Comput. Graph. **44**, 1–10 (2014)
7. dell'Erba, R.: Swarm robotics and complex behaviour of continuum material. Continuum Mech. Thermodyn. **31**(4), 989–1014 (2018). https://doi.org/10.1007/s00161-018-0675-1
8. dell'Erba, R.: Position-based dynamic of a particle system: a configurable algorithm to describe complex behaviour of continuum material starting from swarm robotics. Continuum Mech. Thermodyn. **30**(5), 1069–1090 (2018). https://doi.org/10.1007/s00161-018-0663-5

9. dell'Isola, F., Seppecher, P., Della Corte, A.: The postulations á la D'Alembert and á la Cauchy for higher gradient continuum theories are equivalent: a review of existing results. Proc. R. Soc. A: Math., Phys. Eng. Sci. **471**(2183), 20150415 (2015)
10. Dell'Isola, F., Madeo, A., Seppecher, P.: Cauchy tetrahedron argument applied to higher contact interactions. Arch. Ration. Mech. Anal. **219**(3), 1305–1341 (2016)
11. Abali, B.E., Müller, W.H., dell'Isola, F.: Theory and computation of higher gradient elasticity theories based on action principles. Arch. Appl. Mech. **87**(9), 1495–1510 (2017). https://doi.org/10.1007/s00419-017-1266-5
12. Alibert, J.-J., Seppecher, P., Dell'isola, F.: Truss modular beams with deformation energy depending on higher displacement gradients. Math. Mech. Solids **8**(1), 51–73 (2003)
13. dell'Erba, R.: Rules governing swarm robot in continuum mechanics. Mathematics and Mechanics of Solids. 10812865221088796 (2022). https://doi.org/10.1177/10812865221088795
14. D'Avanzo, P., Rapisarda, A.C., Sirletti, S.S.: From the swarm robotics to material deformations. In: Marmo, F., Sessa, S., Barchiesi, E., Spagnuolo, M. (eds.) Mathematical Applications in Continuum and Structural Mechanics, pp. 87–125. Cham, Springer International Publishing (2021). https://doi.org/10.1007/978-3-030-42707-8_6

Check for updates

Mechanism Design Optimization of a Portable Scanner for Measuring Atmosphere Light Pollution

Eusebio E. Hernández[1(✉)], S. Ivvan Valdez[2], and Hector Lamphar[2]

[1] ESIME Ticomán, Instituto Politécnico Nacional, Miguel Othón de Mendizabal SN, Del. Gustavo A. Madero, Ciudad de México 07320, Mexico
euhernandezm@ipn.mx

[2] CONACYT-Centro de Investigación en Ciencias de Información Geoespacial AC, Parque Tecnológico San Fandila SN, Pedro Escobedo, Querétaro, Mexico
{svaldez,hsolano}@centrogeo.edu.mx

Abstract. In contrast with other kinds of atmospheric pollutants, light pollution does not present highly-variable patterns in short time periods. Hence, the cost of a monitoring network can be avoided, under the assumption that light pollution is almost constant from one day to the next, by using a portable monitoring device. In addition, with an adequate design, such device could be used to acquire measurements in a wide set of of angles providing a full-picture of the sky pollution, such devices are named as sky scanners. In this article, we propose to automatically synthesize dimensions of a four-bar linkage mechanism that best fits the scanning task via an optimization problem, solved via a version of the Boltzmann univariate marginal distribution algorithm (BUMDA). The BUMDA proposes configurations with different lengths and reference positions of four-bar linkage mechanisms; to minimize the distance between points in the actual and desired paths for each candidate configuration. The objective function value is the sum of such distances subject to Grashof constraints; thus, the optimal design produces the minimum distance to the desired path. This proposal for automated design reduces the working time and experience requirements of a human designer, and trial-and-error design intends, by determining adequate dimensions for the mechatronic system. A CAD model and a simulation demonstrate the design feasibility and the high accuracy of the resulting device.

Keywords: Atmosphere light pollution · Portable sky scanner · Mechanism optimization

© The Author(s), under exclusive license to Springer Nature Switzerland AG 2023
M. Ceccarelli et al. (Eds.): TORVEASTRO 2023, MMS 130, pp. 160–176, 2023.
https://doi.org/10.1007/978-3-031-28447-2_13

1 Introduction

Even though night-time outdoor lighting is essential in modern urban areas for safety, recreation, and decorating purposes, it can led to a huge waste of energy and adverse effects on humans and animals. The alteration of the natural quantity of light as a consequence of poorly designed lighting systems and over-illumination of nocturnal environments has been defined as light pollution, and also should be included in the same environmental risk as other environmental pollution problems [23]. Thus, over the last two decades have grown the scientific community interest to quantify light pollution and its effects, in particular those on night-sky brightness and stellar visibility, which is driven by the effort to preserve darkness of the night-sky and the health of the beings. Light pollution is mostly located at the troposphere level, up to 10 km above sea level. At this level, the wasteful light emitted directly by artificial lighting or by reflections from the ground can be scattered by clouds, fog, and pollutants, interacts with suspended particles (specifically aerosols and water molecules), which are the main modulators of light pollution. The conditions of these particles can exhibit extreme variations that make this environment threat an overly complex phenomenon to study. As a consequence of these variations, light pollution values show temporal changes that also occur at different measurement angles. Typically, measurements at the zenith never show the same values as measurements close to horizon. Light pollution is often measured with photometers located at astronomical observatories that offer only local measurements. Having not many portable instruments has limited the efforts of many scientists to obtain data covering larger areas. In addition, one direct consequence of light pollution is a less number of stars visible by unaided eyes, therefore, the magnitude used in astronomy to quantify light pollution is the luminance of the night sky background [23,24].

The measurement of light pollution should be carried out in night environments, when the amount of light is extremely low, and therefore, instruments with high sensitivity are required. There are some instruments photometrically calibrated consisted of CCD (Charge Coupled Device) [1] or CMOS (complementary metal-oxide-semiconductor) sensors with different filters and fisheye lens, which provides a scientific instrument with a 180° field of view in all directions. However, they could be either expensive or complex, and generally installed on permanent observatories. Thus, the development of low-cost devices for measurements of angular daylight luminance distributions is increasing in interest. In [27], a device for the measurement of diffuse solar irradiance on tilted surfaces pointing-to as well as fixed on the equatorial direction is proposed. In [7], a High Dynamic Range image-based system was applied to capture luminance distributions of the sky, window view, and interior surfaces. It has been proposed to predict indoor daylight distribution and optimizing the building's design. More recently, a portable radiometer for measuring the spectral sky radiance linearly over 5–6 orders of magnitud was introduced in [11], which is well suited for narrow-band photometry in both dark areas and bright urban and suburban environments. In [17], a low-cost prototype based on Light Dependent Resistors to measure angular daylight luminance distribution is developed. Although most of these instruments fulfill the applications and they are inexpensive with low maintenance costs, they are adjusted on site by one single operator. Thus, the system's effects on the luminance measurement with a more robust, continuous and automatized operation could be investigated. In particular, a very important point to decide on the best observational strategy to estimate, in real time, the stability of the atmosphere at a certain site and the location of the darkest and the most transparent areas in the sky [1], which makes more relevant the development of a fully portable, but robust at the same time, sky scanner system for measuring light pollution. Traditional sky scanner instruments consist of tracking mechanism or moving head to follow a set of programmed discrete points for measuring data of the sky radiance and luminance across the whole sky. For instance, Fig. 1 shows two tracking systems intended to capture the angular sky luminance distribution at various time-points, a sky scanner developed in the Slovak Academy of Sciences is shown on the left, where movements are achieved by using a traditional 3 DOF mechanism; on the right, the preliminary design of a tracker based on four bar mechanism is presented. It has been established that about one hundred points, evenly distributed across the sky can be recorded during each scan [6]. Thus, a problem with this kind of measurements is the instrument response time for mapping the whole sky since atmosphere may change between two consecutive measurements at different viewing angles.

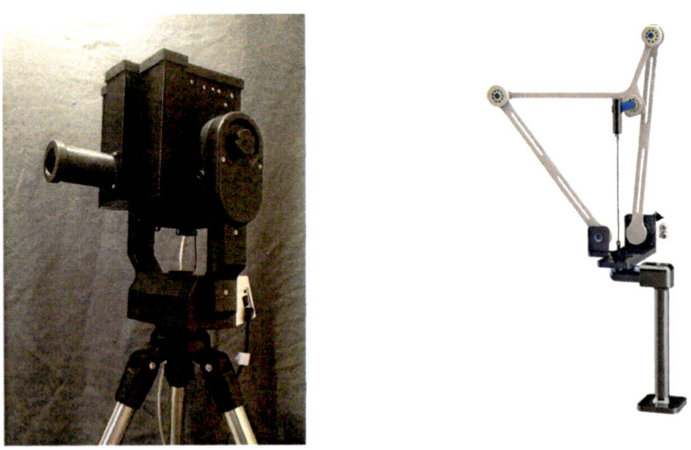

Fig. 1. Left image: sky scanner based on the commercial tracker iPANO [12]. Right image: preliminary design of a tracker based on the use of a parallel mechanism [25].

Over last decades, parallel mechanisms have been proposed to tracking predefined trajectories or tasks in an extensive and diverse range of applications, including those from medicine, manufacturing, packaging, and space utilities. Moreover, the wide range of applications of parallel mechanisms still growing nowadays because some of their inherent and appealing characteristics, namely a blend of speed, precision, and payload capacity that make possible to meet the needs of a variety of those industries. In particular, the generation of controlled movements through planar mechanisms has played an essential role in successful applications into the robotic and mechatronic areas. Indeed, modifications of traditional planar mechanisms have still been proposed to achieve more flexibility, and high accuracy in function generation tasks [22]. Repeatability and automation of predefined tasks are the main advantages of these planar structures, which have been widely used in a diversity of applications, as those for automotive industry [32] or biomechanics [10]. The continuous increase in computer power has permitted approaches for addressing demanding optimization problems, as is the synthesis of robotic mechanisms. This task is a complex process for determining many parameters of mechanism to achieve the required motion or trajectory on its end effector. Traditionally, the synthesis problem for linkage mechanisms has been divided into 3 sub-problems: path, function, and motion generation [8]. The four-bar linkage has been widely used as a case study for optimal synthesis of mechanisms and mechatronic systems [2,3,33] due to its flexibility to reach a wide variety of cyclic paths, and its simplicity. More recently, with the fast development of high-speed computers, the approaches for trajectory generation problem based on metaheuristic optimization have been growing including those using traditional genetic algorithms [5], hybrid strategies [14,16], novel algorithms [9] or complete software packages [28]. Some of these works have the purpose of comparing the performance of the proposed algorithms in terms of the number of function evaluations, computing time, and differences between the target curve and trajectory generated. At this stage, these characteristics can not be considered as the most critical for an optimized design since they do not substantially contribute to improving a build prototype.

1.1 Significance for Aerospace Sciences

Light pollution is triggered by an excess of artificial light at night that interacts with atmospheric particles, producing a nocturnal glow that has multiple environmental effects. In their first approaches, studies on light pollution often focused on astronomical and atmospheric aspects. In fact, one of the main reasons that have historically led astronomical observatories to change their headquarters is the search for cleaner skies, away from the visual disturbances that light pollution produces. However, studies on the consequences of excessive artificial lighting should not only analyze the processes that affect astronomy, since they increasingly reveal the impact that an anomalous amount of light during the night has on aerospace sciences. In this context, there are also studies that try to determine the influence that the atmosphere has on the resulting light pollution. Considering that part of the radiation emitted by the light source escapes to the atmosphere and is absorbed and scattered in the troposphere, mainly because of the presence of aerosols and water molecules, the degree of light pollution is directly proportional to the atmospheric conditions of the space in where the emission occurs. Even assuming that the lighting system of cities is homogeneous in its process of technification, the existence of differences in the atmospheric space produces different lighting results in terms of the amount and type of light pollution generated. Reliable methods for retrieving experimental data are important given the complexity of the problem. In this context, one important aerospace application that light pollution has is the gathering of data from satellites or even from the International Space Station (ISS) [13]. For instance, light pollution researchers often use as data the night-time images captured by the astronauts on the ISS since they represent a valuable source of information [18]. There is a significant advantage in using remote sensing data for satellites or from the ISS, because local measurements require too much time to cover a large area [20,21]. However, one of the major difficulties when analyzing data from the ISS is to locate the right image for the correct photometric study, as many of the images are daytime images and therefore they must be discarded. Also, the database contains images without classification since the ISS does not have a Global Positioning System. In this context, our proposal of a portable scanner for measuring atmospheric light pollution can, in fact, be a valuable source of experimental retrieval from the ISS. The major problem that the astronauts have when taking pictures from the ISS is that they cannot choose a standard window size and type. Normally they take the pictures from the first available window, but at the end the result will have a noise (related to the thickness and type of glass) that later cannot be processed. In this regard, it will be better to avoid any disturbances experienced inside the ISS and have a device for the automatic retrieval of light pollution. Our proposal here will be to install a portable scanner outside the ISS that can be prepared with all the sensors needed and automated to measure light pollution according to the ISS orbit.

In this paper, the automated synthesis of a four-bar mechanism to positioning a portable sky scanner is proposed. An optimization problem is formulated by considering the discretized workspace that must cover the linkage end effector, which has been intended for measuring light pollution. This formulation is implemented and solved with a well established estimation of distribution algorithm that delivers a set of design variables required for CAD modeling of the resulting optimized redesign.

2 Description of Four-Bar Mechanism and Discretized Workspace

The four bar linkage mechanism has a single actuated element, and the end-effector can be posed in non-intuitive positions by manipulating the mechanism's lengths and reference positions, see Fig. 2. The inverse kinematic problem belong two solutions that are solved with a minimum computational effort. All these features make it excellent as a base design for automated synthesis. In the same regard, the portable sky scanner requires a mechanism that can repeat cyclic paths, hence the design task makes a perfect match with the proposed four-bar linkage base design. On the right of Fig. 2 one can see a semicircular path predefined by a set of precision points that should follow the task device. This section discusses the kinematics of the four-bar linkage. It is a pure algebraic model that must be solved per each candidate solution proposed by the optimization algorithm. Solving the model is, actually, to numerically simulate a position of the end effector given an angle in the actuated element.

2.1 Inverse Kinematics

As mentioned above, the inverse kinematic problem consists in determining the position of the end effector C, known as coupler point, given the actuated angular position θ_2.

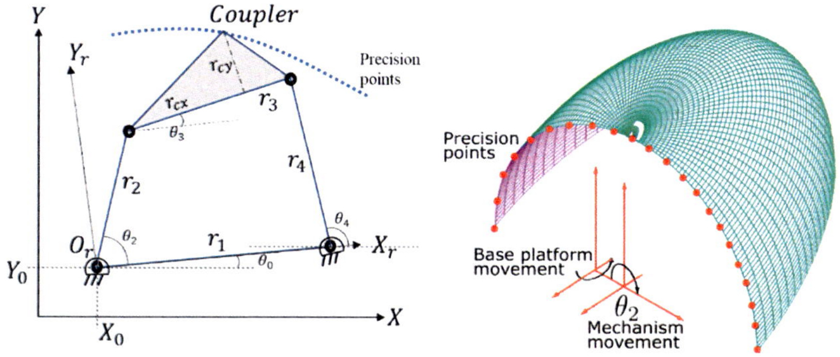

Fig. 2. Left image: representative schematic diagram of a four bar mechanism. Right image: semicircular path predefined by a set of precision points.

The position of this point, that describes the desired path when the crank moves, can be defined in vector notation as

$$\boldsymbol{C_0} = \boldsymbol{r_2} + \boldsymbol{r_{cx}} + \boldsymbol{r_{cy}} \tag{1}$$

These variables are in a local coordinate system O_r, by decomposing Eq. (1) we can obtain

$$\begin{aligned} C_{0x} &= r_2 \cos\theta_2 + r_{cx}\cos\theta_3 - r_{cy}\sin\theta_3 \\ C_{0y} &= r_2 \sin\theta_2 + r_{cx}\sin\theta_3 + r_{cy}\cos\theta_3 \end{aligned} \tag{2}$$

In order to compute θ_3 one can consider the closed-loop vectorial equation of the mechanism

$$\boldsymbol{r_2} + \boldsymbol{r_3} = \boldsymbol{r_1} + \boldsymbol{r_4}. \tag{3}$$

Now Eq. (3) can be outlined in terms of its components around the relative coordinates as

$$\begin{aligned} &r_2\cos\theta_2 + r_3\cos\theta_3 - r_4\cos\theta_4 - r_1 = 0, \\ &r_2\sin\theta_2 + r_3\sin\theta_3 - r_4\sin\theta_4 = 0. \end{aligned} \tag{4}$$

Therefore, Eqs. (3)–(4) can be used to eliminate θ_4, by yielding a simple homogeneous formulation defined as follows

$$k_1 \cos\theta_3 + k_2 \cos\theta_2 + k_3 = \cos(\theta_2 - \theta_3). \tag{5}$$

In this way, θ_3 can be determined as

$$\theta_3 = 2\arctan\left(\frac{-b \pm \sqrt{b^2 - 4ac}}{2a}\right). \tag{6}$$

where

$$\begin{aligned} a &= \cos(\theta_2) - k_1 + k_2 \cos(\theta_2) + k_3, \\ b &= -2\sin(\theta_2), \\ c &= k_1 + (k_2 - 1)\cos(\theta_2) + k_3, \\ k_1 &= r_1/r_2, \\ k_2 &= r_1/r_3, \\ k_3 &= (r_4^2 - r_1^2 - r_2^2 - r_3^2)/(2r_2r_3). \end{aligned} \tag{7}$$

Finally, the position the coupler C, defined in terms of the general coordinates can be expressed as

$$\begin{aligned} C_x = X_0 &+ r_2 \cos(\theta_0 + \theta_2) + r_{cx} \cos(\theta_0 + \theta_3) \\ &- r_{cy} \sin(\theta_0 + \theta_3), \end{aligned} \tag{8}$$

$$\begin{aligned} C_y = Y_0 &+ r_2 \sin(\theta_0 + \theta_2) + r_{cx} \sin(\theta_0 + \theta_3) \\ &+ r_{cy} \cos(\theta_0 + \theta_3). \end{aligned} \tag{9}$$

Thus, one can locate the coupler point C in a global coordinate system, through X_0, Y_0, and θ_0. Equations (8) and (9) are used in the objective function, in order to determine the quality for a given configuration. One can note that Eq. (6) has three kinds of solutions, two of them real, and the others are complex. The two real solutions are valid configurations, named as crossed and open configurations, but the complex is not valid. It is worth noting that these equations can belong real and complex solutions for the same lengths and reference positions, this means that for some values of θ_2 the configuration is valid, but for others it is not.

3 The Boltzman Univariate Marginal Distribution Algorithm (BUMDA)

Estimation of Distribution Algorithms (EDAs) constitute a kind of metaheuristic algorithm [4,15,19] derived from the probabilistic modeling of Genetic Algorithms.

EDAs generate offsprings by sampling from a probabilistic model, estimated from the parent set. In this regard we use a version of the Boltzman Univariate Marginal Distribution Algorithm (BUMDA) [31] for the optimization task. In the BUMDA the probabilistic model is a Gaussian distribution, which is estimated to approximate the Boltzman distribution which energy function is the objective function to maximize (minus the objective function to minimize).

The BUMDA is a real-coded EDA that assumes independence of the decision variables, nevertheless, it actually have solved multivariate problems according to [31].

The functions $\mathcal{P}_{k,i}(\theta^i_{k,h})$ are normal probability functions with mean $\mu_{k,i}$ and variance $\sigma^2_{k,i}$, for $0 < i \leq l$ variables. Each probability function is estimated and sampled independently. In Eq. (10) k is the iteration or generation counter, i is the number of variable, $\theta^i_{k,s}$ is the i-positioned value of the s candidate solution and $p_{k,i}(\theta^i_{k,s}|S_{\theta_k})$ are weights associated with each of the N_{ind} selected individuals.

$$\mu_{k,i} = \sum_{s=1}^{N_{ind}} \theta^i_{k,s} p_{k,i}(\theta^i_{k,s}), \quad \sigma^2_{k,i} = \sum_{s=1}^{N_{ind}} (\mu_{k,i} - \theta^i_{k,s})^2 p_{k,i}(\theta^i_{k,s}), \tag{10}$$

The weighs $p_{k,i}(\theta^i_{k,s}|S_{\theta_k})$ in Eq. (10) are computed in each iteration k, using the selected set from the truncation method [31]. The selected set are those individuals with objective function value equal or better than the threshold ν_k. In the first generation, the threshold ν_0 is the worst objective function value in the initial population, from the second to the last generation the next threshold ν_k, is computed using the sorted objective function values in the population as follows: if the median of the objective values in the population is better than ν_{k-1}, then ν_k is set to this median, otherwise it is the first objective function value in the population that is better than $\nu_{k-1} - 10^{-8}$, that is to say, it is slightly better than the last threshold, hence, this ensures convergence to the best individual. The assigned weight to each $\theta^i_{k,s}$ in the iteration k, for $0 < i \leq l$ and $\vec{\theta}_{k,s} \in \theta_k$, is given by Eq. (11). Where $g(\cdot) = f(\cdot) - f_{k,min} + 10^{-8}$, where $f_{k,min}$ is the minimum objective function value in the population at generation k. The function $g(\cdot)$ translates the objective function values to make them all positive and greater than 10^{-8} except to the worst that is exactly 10^{-8}, thus the worst individual gets a very small weight in the parameter computation.

$$p_{k,i}(\theta^i_{k,s}|S_{\theta_k}) = \begin{cases} \frac{g(\vec{\theta}_{k,s})}{\sum_{s=1}^{N_{ind}} g(\vec{\theta}_{k,s})}, & \text{if } \vec{\theta}_{k,s} \in S_{\theta_k}, \\ 0, & \text{if } \vec{\theta}_{k,s} \notin S_{\theta_k}. \end{cases} \tag{11}$$

In this research, according to Fig. 2 the *BUMDA* is applied to estimate the parameters $\vec{\theta} = \{\theta_0, r_1, r_2, r_3, r_4, r_{cx}, r_{cy}, X_0, Y_0)\}$.

3.1 BUMDA Modifications

In contrast to the original BUMDA, in this work we propose three modifications that benefit the algorithm performance, the most important is the inclusion

Algorithm 1 *BUMDA*

1: Input: N_{ind}, N_{elite}, N_{gen}, tol.
2: Generate an initial population θ_0 with N_{ind} individuals.
3: $k = 0$
4: $sdf = tol + 1$
5: **while** $k < N_{gen}$ and $count \leq 5$ and $sdf > tol$ **do**
6: Evaluate θ_k on the objective function.
7: Select the best individuals, $S_{\theta_k} \subset \theta_k$, by truncating objective function values below ν_k.
8: n_s = number of selected individuals.
9: $n = min(N_{elite}, n_s)$
10: **if** $f^{sorted}_{k-1,n} + 10^{-8} < f^{sorted}_{k,n}$ **then**
11: $count = 0$
12: **else**
13: $count = count + 1$
14: **end if**
15: Estimate the probability distribution function $P(x|S_{\theta_k})$, see Eq. (10).
16: Sample θ_{k+1} with $N_{ind} - n$ new individuals from $P(x|S_{\theta_k})$.
17: Concatenate the current best n individuals with the sampled in the step above.
18: $sdf = stddev(f(\theta_k))$
19: $k = k + 1$.
20: **end while**
21: Return the best solution as the solution to the maximization problem.

of the n_{elite} individuals that are preserved. In this version we can preserve 1 or more of the best individuals, this simple modification changes the way the exploration and exploitation is managed by the BUMDA. The original algorithm preserves *a single elite individual* while our version preserve at most the half of the population but the number of preserved individuals, denoted as n_{elite} is a user parameter. Hence, promissory individuals in different regions could be preserved favoring an intensive exploration. On the other hand, if all the n_{elite} individuals are in the same region the rapid BUMDA convergence is not affected. A second modification is the stopping criteria, we use the standard deviation of the objective function of the population instead of the variables because they are highly related (in continuous functions a small variance in the variables implies a small variance in the objective function values), but it is computationally cheaper than computing the standard deviation of the objective function values. Another modification to the stopping criteria, is the counter criteria, the counter in line 11 of Algorithm 1, is increased if the worst elite individual does not improve in 5 consecutive generations. This is a way of measuring stagnation, notice that testing for improving the last elite individuals actually is a way of testing if any individual in the elite set improves, additionally if any of the elite individuals improves during 5 generations, then it is unlikely that it will improve during additional generations, because the probability function depends on those individuals, and if those individuals do not change the probability distribution does not change either, and if the probability distribution is incapable of sampling a single slightly better elite individual in, usually, some hundreds of independent samples, then it is unlikely to sample a better individual in additional samples.

3.2 Parameter Settings

We use a population size 270 (30 times the number of decision variables). The inferior search limits are $[0, 0.1, 0.1, 0.1, 0.1, 0.01, 0.01, -0.5, -0.5]$, the superior search limits are $[2\pi, 1, 1, 1, 1,\ 1, 1, 1.5, 1.5]$, for $\vec{\theta} = \{\theta_0, r_1, r_2, r_3, r_4, r_{cx}, r_{cy}, X_0, Y_0)\}$, respectively.

Even though it is not in the scope the parameter settings analysis. This setting where sound via trial-error executions, we do no perform a rigorous search because they are intensive in computational time, but we can argue that the first limit covers the full range of possible values of the variable for the reference angle θ_0, the inferior and superior limits of the lengths are set according physical limitations, a length lower than 10 cm could require a high torque and presents a smaller dexterity, while a length greater than a meter makes the device less portable. Finally, in our final design there are not negative positions (X_0, Y_0), but we notice through several executions that setting the inferior search limits to -0.5 deliver better results, that is to say, the BUMDA performs better if the search limits of this two variables is larger than it needs.

4 Automated Synthesis Problem Definition

As was presented before in Fig. 2, the portable sky scanner must cover a kind of half sphere. Thus, the discretized workspace is represented by a quarter of such sphere, by showing a cut, it defines a predefined path for the four-bar linkage mechanism. Actually, covering the surface is not as important as covering the zenith and azimuth angles (Fig. 3). The first is achieved via the four-bar linkage mechanism, while the second is covered by a moving plate.

This work deals with path synthesis, which is intended to minimize the difference between the desired path and the one described by the coupler point of the linkage. Thus, the optimization problem is to find the mechanism that reaches with minimum error a set of points named *precision points*. Hence, the optimization model is in Eq. (12). The objective function sums the square difference between the N precision points, labeled as $[C^P_{x,i}, C^P_{y,i}]$, and their closest points in the mechanism's path $[C^M_{x,i}, C^M_{y,i}]$.

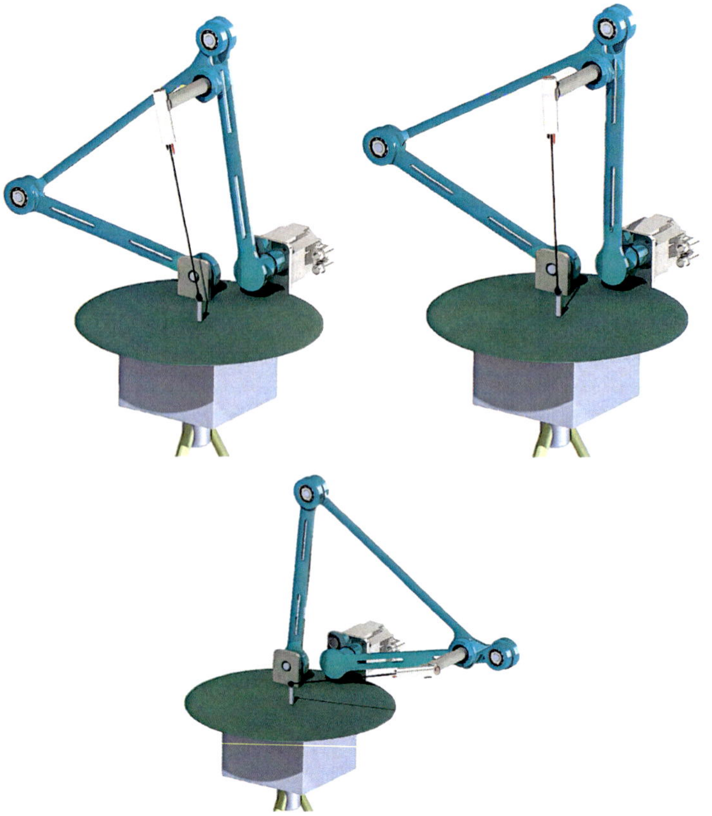

Fig. 3. Different positions of the optimized redesign for scanning in the zenith angle.

$$\min f(\theta_0, r_1, r_2, r_3, r_4, r_{cx}, r_{cy}, X_0, Y_0) = \sum_{i=1}^{N} (C_{x,i}^P - C_{x,i}^M)^2 + (C_{x,i}^P - C_{y,i}^M)^2 \tag{12}$$

$$\begin{aligned} 0 \leq \theta_0 \leq 2\pi \\ 0.1 \leq r_1, r_2, r_3, r_4, r_{cx}, r_{cy} \leq 0.5 \\ 0 \leq X_0, Y_0 \leq 1 \end{aligned} \tag{13}$$

$$\begin{aligned} r_1 + r_3 > r_2 + r_4 \\ r_1 + r_4 > r_2 + r_3 \\ r_3 + r_4 > r_1 + r_2. \end{aligned} \tag{14}$$

4.1 Evaluation of a Candidate Configuration

The optimization algorithm proposes a set of candidate solutions, each of them consist of an array of 9 elements $[\theta_0, r_1, r_2, r_3, r_4, r_{cx}, r_{cy}, X_0, Y_0]$, and it must be evaluated in the objective function in (12) within the valid ranges of the set constraints (13). The set (14) ensure that the resulting mechanism satisfy Grashof constraint. Then, the angular positions of θ_2 (see Fig. 2), are discretized in the range $[0, 2\pi]$, with intervals of $\pi/18000$, that is to say, the end effector position is computed for different positions of θ_2, in its complete working range, displacing $0.01\,^\circ$C the angle between consecutive computations. We associate an error array of size N to the set of N precision points. It is initialized with a large value before starting the simulation. For each value of θ_2 we verify whether the current position of the end effector has a lower distance to any of the precision points. All the precision points with a lower distance to the current end-effector position, require to update the corresponding positions in the error array.

Notice the following:

- Some configurations can not be simulated, because the dimensions of the elements can not build a working mechanism. In this case, the error array remains with a set of large values, thus their sum, that is larger, is returned.
- If a single value of θ_2 delivers a real value of the end-effector's position, all the values in the error array are modified.
- If a subset of values of θ_2 deliver real values of the end-effector's position, this is sufficient to compute the objective function, and possible, to deliver a competitive configuration. Hence, it is not necessary that the optimum configuration delivers a real solution to all the θ_2 values, neither in the optimization process, nor in the physical mechanism.

4.2 Optimization Method

The optimization method is the Boltzmann Univariate Marginal Distribution Algorithm (BUMDA) [26,29–31]. It is an Estimation of Distribution Algorithm based on the Boltzmann distribution approximated by a normal distribution. The BUMDA generates an initial population, with a uniform distribution within the search limits defined in Eq. (12). Then, each candidate solution is evaluated. The best solutions are selected and weighted proportional to a translation of their objective function value, for computing weighted estimators of the mean and variance of the normal distribution. Then, new samples are generated from the normal distribution and they are used to replace the current population except the elite individual, the process is repeated until convergence is reached. The convergence is measured using the standard deviation of the objective function values of the population. The BUMDA, presents several advantages, for instance, it do not require of an arbitrary stopping criterion, as other population-based algorithms, and it requires a single parameter, that is the population size, and is set to $30n_{var} = 90$, the minimum standard deviation used as stopping criterion is 10^{-8}.

5 Results of the Implemented Optimization

In Table 1, it is shown a typical output of the BUMDA algorithm. It shows the output dimensions of the mechanism and the reference positions and angle performed by the algorithm. The f value denotes the objective function value, that also is an accumulated distance from the desired discretized path to the actual path. These two paths are shown in Fig. 4, the actual path, with red lines, is quite close to the precision points (with blue crosses). Thus these results demonstrate that the method actually design a feasible mechanism.

Finally, Fig. 5 shows a 3D CAD model of the final design, the optimized mechanism is shown in light blue color, the end-effector is modeled as a light-weight element, with the optical sensor mounted as a cylinder with a cable. It varies the orientation of the sensor to follow a normal angle to the semi-circular path described by the four-bar linkage mechanism. The green base is equipped with an actuator and an axial thrust ball bearing to provide rotation movement of the base platform. Due to the simplicity of the resulting mechanism and the light weight of the moving sensor, the material of the elements is proposed to be of rigid polymer, which favors the energy saving in the task of conducting the light sensor. Besides, in each joint mechanism ball bearings are considered to reduce the friction between the rotating links. Two step DC motors are considered to actuate the input crank link and the base platform.

Table 1. Optimized parameters

θ_0	3.7035 rad rad
r_1	0.1000 m m
r_2	0.5391 m m
r_3	0.6663 m m
r_4	0.4797 m m
r_{cx}	0.0884 m m
r_{cy}	0.0255 m m
X_0	-0.0044 m m
Y_0	0.0134 m m
f	$4.94e-5\ m^2$

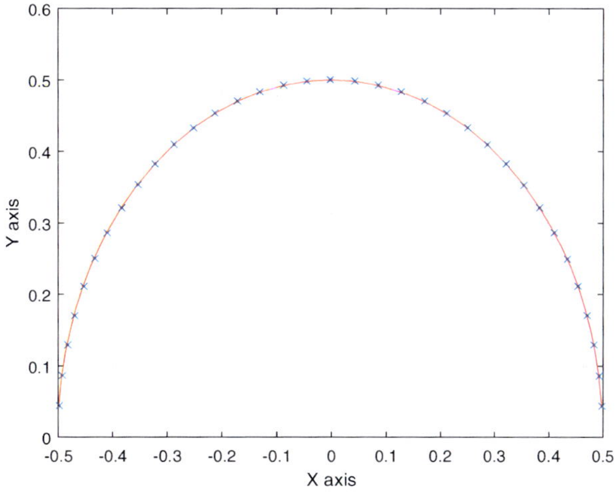

Fig. 4. Approximation of the desired trajectory (red) through the path optimization (precision points).

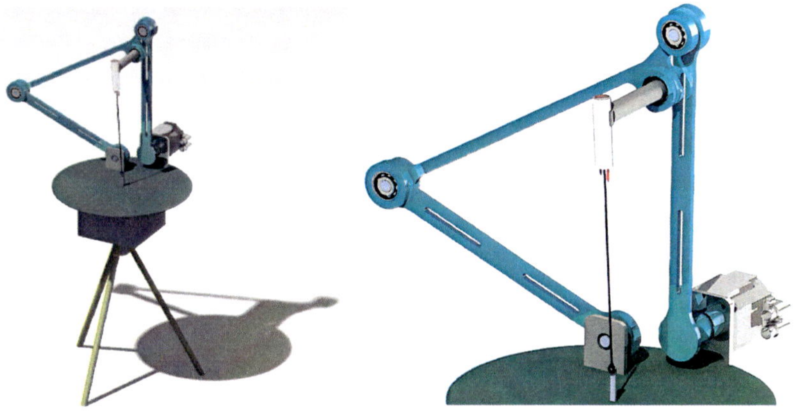

Fig. 5. Left image: a 3D CAD model with optimized mechanism. Right image: Close-up to the sensor mounting Fig. 4.

6 Conclusions

This work presents a novel design of sky scanner mechanism, which dimensions are automatically synthesized via the BUMDA. This system is intended to measure light pollution over the atmosphere in night conditions, with the purpose of delivering a full range set of measures in equidistant angles over a semi-sphere. Even thought several approaches have studied the path synthesis of four-bar mechanisms, in this paper we have taken advantage of the body of knowledge in the state of the art of different problems, for instance, automated design, a bioinspired algorithm that has shown excellent performance in similar problems, simulation and modelling of a mechatronic system, and the proposal of using a portable sky scanner, which has been proposed ad hoc for the task of replacing a network of sensors. In addition, we introduce slight modifications to the BUMDA that, without extensive analysis that must be performed in the future, report adequate results for this particular problem. Notice that this article is focused on the design need of the sky scanner, without a deep regard of the computational and mathematical tools used for the task. For the formulated optimization problem, the mechanism coupler point must follow a semicircular arc in a sequence of angles, the accumulated error of the optimized design is $4.94e - 5$. It is worth noting this value is the sum of the squares of the error, then the error for each point is <0.001 and it can be considered sufficient for the application. Moreover, it can be reduced in several manners, for instance, by increasing the number of precision points, increasing the size of the discretization of θ_2, and reducing the standard deviation of the objective function values, that is used as the stopping criterion in the optimization algorithm.

At this stage, one of the main objective is to synthesize an optimized design for accomplishing the sky scanner path, and then we have not considered the comparison between different bio-inspired algorithms. It is possible that another algorithm performs better than the BUMDA, but this is not relevant because the error delivered by the algorithm is not significant in a physical application. Additionally, assuming that other algorithm reports a lower error, lengths or dimensions with more decimals below the tenth-thousands, are not physically noticeable for the building process. Our research problem is to model a real-world device via an optimization problem that is solved by a bio-inspired algorithm. In this sense, future work includes additional design specifications to develop an integral design, for instance, to add dexterity, rigidity, or other kinds of constraints. From the point of view of automated design of the sky scanner mechanism, we can improve the optimization model, thus, future work considers including a control scheme and dexterity constraints, even including them in a mechatronic design model.

Acknowledgments. S. Ivvan Valdez is supported by Cátedra-CONACYT 7795. Authors are also grateful to SIP-IPN for supporting part of this work through project SIP-20220607. H. Lamphar is supported by the CONACYT project 376106, by Cátedras Conacyt and by the Slovak Research and Development Agency under Project No. APVV-18-0014.

References

1. Aceituno, J., Sánchez, S.F., Aceituno, F.J., Galadí-Enríquez, D., Negro, J.J., Soriguer, R.C., Gomez, G.S.: An all-sky transmission monitor: Astmon. Publ. Astronom. Soc. Pac. **123** (2011). https://doi.org/10.1086/661918
2. Acevedo, M., Orvananos, T., Velazquez, R., Haro, E.: Optimum balancing of the four-bar linkage using fully cartesian coordinates. IEEE Latin Am. Trans. **17** (2019). https://doi.org/10.1109/TLA.2019.8896821
3. Acharyya, S., Mandal, M.: Performance of EAS for four-bar linkage synthesis. Mech. Mach. Theory **44**(9), 1784–1794 (2009)

4. Bosman, P.A.N., Thierens, D.: Linkage information processing in distribution estimation algorithms. In: Proceedings of the Genetic and Evolutionary Computation Conference GECCO-99 1, pp. 60–67 (1999)
5. Cabrera, J., Simon, A., Prado, M.: Optimal synthesis of mechanisms with genetic algorithms. Mech. Mach. Theory **37**(10), 1165–1177 (2002). https://doi.org/10.1016/S0094-114X(02)00051-4
6. Chauvin, R., Nou, J., Thil, S., Grieu, S.: Modelling the clear-sky intensity distribution using a sky imager. Solar Energy **119** (2015). https://doi.org/10.1016/j.solener.2015.06.026
7. Chiou, Y.S., Huang, P.C.: An HDRi-based data acquisition system for the exterior luminous environment in the daylight simulation model. Solar Energy **111** (2015). https://doi.org/10.1016/j.solener.2014.10.032
8. Erdman, A.G.: Computer-aided mechanism design: now and the future. J. Mech. Des. Trans. ASME **117** (1995). https://doi.org/10.1115/1.2836476
9. Huang, Q., Yu, Y., Zhang, K., Li, S., Lu, H., Li, J., Zhang, A., Mei, T.: Optimal synthesis of mechanisms using repellency evolutionary algorithm. Knowl.-Based Syst. **239** (2022). https://doi.org/10.1016/j.knosys.2021.107928
10. Kittisares, S., Nabae, H., Endo, G., Suzumori, K., Sakurai, R.: Design of knee support device based on four-bar linkage and hydraulic artificial muscle. ROBOMECH J. **7**(1), 1–10 (2020). https://doi.org/10.1186/s40648-020-00165-2
11. Kocifaj, M., Kómar, L., Kundracik, F.: Pepss—a portable sky scanner for measuring extremely low night-sky brightness. J. Quant. Spectrosc. Radiat. Trans. **210** (2018). https://doi.org/10.1016/j.jqsrt.2018.02.017
12. Kundracik, F., Kocifaj, M.: Whole-sky scanner of the light pollution. www.davinci.fmph.uniba.sk/~kundracik1/SkyScanner/indexe.html
13. Kyba, C., Aubé, M., Bara, S., Bertolo, A., Bouroussis, C., Cavazzani, S., Espey, B., Falchi, F., Gyuk, G., Jechow, A., Kocifaj, M., Kolláth, Z., Lamphar, H., Levin, N., Liu, S., Miller, S., Ortolani, S., Pun, C.S., Ribas, S., Zhang, J.: The benefit of multiple angle observations for visible band remote sensing using night lights (07, 2021). https://doi.org/10.1002/essoar.10507575.1
14. Laribi, M.A., Mlika, A., Romdhane, L., Zeghloul, S.: A combined genetic algorithm-fuzzy logic method (GA-FL) in mechanisms synthesis. Mech. Mach. Theory **39** (2004). https://doi.org/10.1016/j.mechmachtheory.2004.02.004
15. Larrañaga, P., Etxeberria, R., Lozano, J.A., Peña, J.M.: Combinatorial optimization by learning and simulation of Bayesian networks. In: Boutilier, C., Goldszmidt, M. (eds.) Uncertainty in Artificial Intelligence, UAI-2000 1, pp. 343–352 (2000a)
16. Lin, W.Y.: A GA-DE hybrid evolutionary algorithm for path synthesis of four-bar linkage. Mech. Mach. Theory **45** (2010). https://doi.org/10.1016/j.mechmachtheory.2010.03.011
17. Maskarenj, M., Banerjee, R., Ghosh, P.C.: Design and development of a low-cost angular sky luminance measurement system. Build. Environ. **142**, 22–33 (2018). https://doi.org/10.1016/j.buildenv.2018.06.001
18. Sanchez de Miguel, A., Zamorano, J., Pascual, S., López-Cayuela, M.Á., Ocaña, F., Challupner, P., Gomez Castaño, J., Fernández-Renau, A., Gómez, J., Miguel, E.: ISS nocturnal images as a scientific tool against light pollution: flux calibration and colors, pp. 916–919 (05, 2013)
19. Mühlenbein, H., Paaß, G.: From recombination of genes to the estimation of distributions I. Binary parameters. In: Voigt, H.M., Ebeling, W., Rechenberg, I., Schwefel, H.P. (eds.) Parallel Problem Solving from Nature—PPSN IV, pp. 178–187. Springer, Berlin Heidelberg (1996)

20. Pack, D.W., Coffman, C., Santiago, J.R.: A year in space for the cubesat multi-spectral observing system: Cumulos. Slides SSC19 (2019)
21. Pack, D.W., Hardy, B.S., Longcore, T.: Studying the earth at night from cubesats. In: Proceedings of the AIAA/USU Conference on Small Satellites, Session 10: Leo Missons SSC17 (2017)
22. Patel, H.R., Mungla, M.J.: Optimal synthesis for function generation of double-loop planer four-bar mechanism using genetic algorithm. SN Appl. Sci. **1**(5), 1–8 (2019). https://doi.org/10.1007/s42452-019-0407-6
23. Pun, C.S.J., So, C.W.: Night-sky brightness monitoring in Hong Kong. Environ. Monitor. Assess. **184**(4), 2537–2557 (2012)
24. Ratledge, D.: The Art and Science of CCD Astronomy. Springer Science & Business Media (2012)
25. Rios, A., Hernández, E.E., Lamphar, H., Valdez, S.I.: Automated dimensional synthesis of a portable sky scanner for measuring light pollution. In: Proceedings of the 2021 Mexican International Conference on Computer Science, ENC 2021 (2021). https://doi.org/10.1109/ENC53357.2021.9534811
26. Ríos, A., Hernández, E.E., Valdez, S.I.: A two-stage mono- and multi-objective method for the optimization of general UPS parallel manipulators. Mathematics **9**(5) (2021). https://doi.org/10.3390/math9050543, www.mdpi.com/2227-7390/9/5/543
27. de Simon-Martin, M., Alonso-Tristan, C., Gonzalez-Peña, D., Diez-Mediavilla, M.: New device for the simultaneous measurement of diffuse solar irradiance on several azimuth and tilting angles. Solar Energy **119**, 370–382 (2015). https://doi.org/10.1016/j.solener.2015.06.001
28. Torres-Moreno, J.L., Cruz, N.C., Álvarez, J.D., Redondo, J.L., Giménez-Fernandez, A.: An open-source tool for path synthesis of four-bar mechanisms. Mech. Mach. Theory **169** (2022). https://doi.org/10.1016/j.mechmachtheory.2021.104604
29. Valdez, I., Herández Aguirre, A., Botello, S., Hernández-Martínez, E.E.: Síntesis óptima de mecanismos utilizando un eda basado en la distribución normal. Nova scientia **6**(11), 45–66 (2014)
30. Valdez, S.I., Botello-Aceves, S., Becerra, H.M., Hernández, E.E.: Comparison between a concurrent and a sequential optimization methodology for serial manipulators using metaheuristics. IEEE Trans. Indus. Inform. **14**(7), 3155–3165 (2018)
31. Valdez, S.I., Hernández, A., Botello, S.: A Boltzmann based estimation of distribution algorithm. Inform. Sci. **236**, 126–137 (2013)
32. Yildiz, A.: Parametric synthesis of two different trunk lid mechanisms for sedan vehicles using population-based optimisation algorithms. Mech. Mach. Theory **156** (2021). https://doi.org/10.1016/j.mechmachtheory.2020.104130
33. Zhang, W., Li, Q., Guo, L.: Integrated design of mechanical structure and control algorithm for a programmable four-bar linkage. IEEE/ASME Trans. Mechatron. **4**(4), 354–362 (1999)

Repetitive Path Planning with Experience-Based Bidirectional RRT

Guoyu Zuo[1,2], Chun Wu[1,2], and Gao Huang[1,2](✉)

[1] Beijing Key Laboratory of Computing Intelligence and Intelligent Systems, Beijing 100124, China
{zuoguoyu,wuchun,huanggao}@bjut.edu.cn

[2] Faculty of Information Technology, Beijing University of Technology, Beijing 100124, China

Abstract. Repetitive motion planning in semi-structured environments is involved in many robotics applications, such as manufacturing and in-orbit service of spatial robotic arms. In these scenarios, the environment is mainly static, but it may contain some movable obstacles that the robot needs to avoid, and the planning problems that the robot will encounter are very similar. However, most current sampling-based motion planning methods ignore the correlation among tasks, and instead plan each task instance from scratch, or assume that the environment is highly structured. In order to overcome this problem, this paper proposes an experience-based bidirectional rapidly-exploring random trees* (EBRRT*) for repetitive path planning in semi-structured environments. First, an adaptive sampler learned from the demonstration paths by a Gaussian mixture model (GMM) is constructed, which can guide the planner to generate samples in task-related regions. Second, an experience graph is proposed to capture the exploration of configuration space by previous similar tasks. Subsequent tasks can retrieve locally valuable information from the graph to avoid exploring the configuration space repeatedly. In addition, the validity of edges contained in the experience graph is lazily verified during the planning process so that the experience graph can adapt to small environmental changes. Finally, the proposed method is verified in a 2D simulation environment and a 7-DOF manipulator scene. Experimental results show that the proposed method can return a high-quality path in less time than other state-of-the-art sample-based motion planning for repetitive path planning in semi-structured environments.

Keywords: Repetitive path planning · Experience graph · Gaussian mixture model (gmm) · Semi-structured environments

---

Supported by the National Nature Science Foundation of China under Grant 61873008, 62103053, and the Nature Scienc Foundation of Beijing under Grant 4192010. Corresponding author is Gao Huang (email: huanggao@bjut.edu.cn).

© The Author(s), under exclusive license to Springer Nature Switzerland AG 2023

M. Ceccarelli et al. (Eds.): TORVEASTRO 2023, MMS 130, pp. 177–192, 2023.
https://doi.org/10.1007/978-3-031-28447-2_14

1 Introduction

Robots have been widely used in real-world applications such as parts assembly, mail sorting in mailrooms, or pick-and-place tasks on conveyor belts [1–3]. Moreover, with space technology development in recent years, more and more spacecrafts are equipped with robotic arms. The spatial robotic arms can effectively replace astronauts to complete various on-orbit services (for example, on-orbit capture, on-orbit maintenance and space debris cleanup) [4–6]. In these scenarios, the obstacles in the environment are usually fixed or static. However, there are also small dynamic changes in the environment due to the robot manipulating objects, which may invalidate the previously calculated regions where the robot can move freely without collision [7]. Such environments are classified as semi-structured. In addition, the manipulation tasks in these scenarios are usually highly repetitive, leading to similar planning problems [8]. A typical case can be seen in Fig. 1, where the robot needs to move all the blue cans from left to right while avoiding obstacles. Most objects in the environment (e.g., table, yellow box, shelf, etc.) can be assumed to be static, but there are still some slight dynamic changes due to attached and moved objects. In addition, each motion planning problem is similar but not identical. Faced with such situations, humans have a great capacity to learn from the past. When they encounter a similar task or environment, they can recall what they learned from the past and respond quickly to the current task. However, robots usually plan each task from scratch even if the tasks are very similar. To achieve human-level motion planning, how to give robots the ability to learn from the past and adapt to environmental changes is the impetus for this paper.

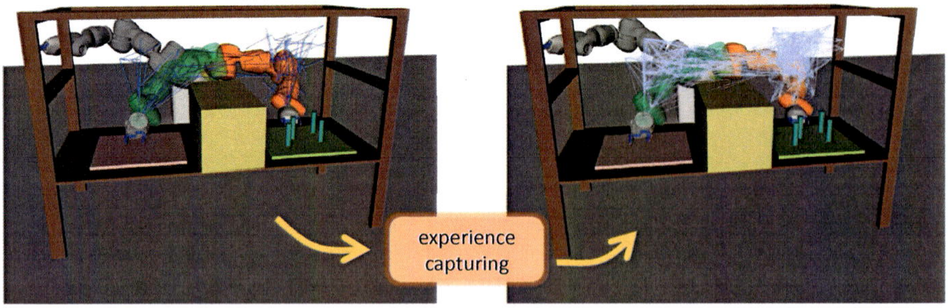

Fig. 1. A robot performs the repetitive task of moving the blue box from the green region to the pink region.

The sampling-based motion planners (SBMPs) are wildly used in practical applications due to their robustness and effectiveness. Among them, roadmap-based motion planning methods such as Probabilistic Roadmaps (PRM) [9] work by randomly sampling in the configuration space and preserving the connectivity of the configuration space in a graph structure. They are suitable for multi-query motion planning problems in highly-structured environments. However, when the environment changes, they need to spend much time verifying the validity of all edges in the graph by collision detection, which is often regarded as the bottleneck of most motion planning algorithms. In lazzyPRM [10], collision detection adopts a delayed detection strategy, making it adapt to small environmental changes. However, building a complete roadmap in advance for all graph-based methods is usually computationally expensive. Sufficient sampling points need to be obtained from the entire configuration space, which contains many areas irrelevant to the planning task. The tree-based methods, like the rapidly-exploring random tree (RRT) [11], begin with a given starting point as the root node of the search tree. They incrementally explore the configuration space by randomly generating leaf nodes to expand the tree structure and return a feasible path when the leaf node closes to the goal point. Based on them, many variants are proposed to improve road quality or reduce computation time. For example, RRT* and PRM* [12] get the locally optimal path with sufficient samples. BiRRT [13] and BiRRT* [14] take the start and goal configurations as the root nodes to build two trees and explore the configuration space at the same time. A common shortcoming of sampling-based motion planning algorithms is the random sampling strategy, which makes them spend much effort sampling in task-irrelevant regions, especially in high-dimensional spaces. Furthermore, in scenarios where the robot needs to pass the "challenging regions" of the search space (e.g., narrow passages [15]), hitting a sample inside these regions is troublesome via a random sampling strategy, which significantly limits the overall performance of these methods in many scenarios.

Several previous works proposed non-uniform sampling strategies for SBMP, resulting in significant performance gains. The authors of [16] proposed Target-biased Informed Trees (TBIT*), which can guide the search tree to grow toward the target direction to find the initial solution quickly; after the initial path is searched, heuristic sampling is used to optimize the initial path. ADD-RRV [17] uses principal component analysis (PCA) and an improved bridge measurement method to identify the characteristics of local configuration space to improve the randomness of uniform sampling in complex environments. BIT* [18] defines the sampling area as a hypersphere to reduce the computational burden. However, the motion planning algorithms mentioned above improve the randomness of uniform sampling to a certain extent but do not transfer the knowledge between different planning queries. They plan each task from scratch without considering the correlation between tasks. The knowledge contained in previous motion plans may provide helpful information to speed up the solution of subsequent planning tasks.

Using the prior knowledge from past motion plans to speed up the solution of subsequent planning problems has attracted the attention of a large number of researchers in recent years. Ichter et al. used a Conditional variational Auto-Encoder (CVAE) to learn the sampling distribution to generate samples in "interesting" regions where the optimal trajectory likely exists [19]. Wang et al. proposed a neural RRT* method, which can predict the probability distribution of the optimal path [20]. Shah et al. used an Unet neural network to identify critical regions for a given motion planning problem and utilized it to bias sampling for sampling-based motion planning algorithms [21]. The learning-based methods have achieved satisfactory results in some specific scenarios. Unfortunately, these methods depend highly on the dataset's quality and require many samples to infer knowledge. Some approaches attempt to extract features from the demonstration paths and then construct a probabilistic roadmap [22,23]. Like most PRM algorithms, a major limitation is a need to predefine the number of nodes in the graph. A large number of nodes led to significant execution times, and the small number had no guarantee of generating a feasible plan. Choosing an appropriate number of nodes is often tricky, especially for motion planning problems in high-dimensional spaces. In contrast, our method builds an experience graph online and adaptively by exploiting the exploration of the configuration space from previous tasks.

In this paper, an experience-based bidirectional rapidly-exploring random trees* (EBRRT*) method is proposed to solve repetitive motion planning problems in semi-structured environments. It inherits the fast-planning properties of existing incremental sampling-based planners and can capture helpful information from previously computed motion plans. The proposed planner contains two modules. One is an adaptive sampler learning from the demonstrated path, which can guide the samples into the task-relevant region. The other is an experience graph used to capture the exploration of configuration space by previous tasks. Furthermore, a lazy collision detection strategy is introduced, making the experience graph adaptable to small environmental changes.

The remainder of this article is organized as follows: Sect. 2 describes the notions used in this paper. Section 3 introduces the detail of the proposed method. Section 4 is the description of the experiments and experimental results. Conclusion is drawn in Sect. 5.

2 Problem Definition

In this manuscript, we are interested in the repetitive path planning for robots in semi-structured environments. This section describes some notions used in this paper.

Given a robot, let the robot's configuration space (*C-space*) be defined as $\mathcal{C} \subseteq \mathbb{R}^d$, which represents the set of all possible configurations of the robot, where d is the dimension of the configuration space and $q \in \mathcal{C}$ is a possible robot configuration (e.g., A group of joint angles). Let $\mathcal{C}_{free} \subset \mathcal{C}$ denote the free configuration space in which the robot can move freely, and $\mathcal{C}_{obs} = \mathcal{C} \backslash \mathcal{C}_{free}$ be the subspace in which the robot will collide with the environment or itself.

Definition 1. A path $\sigma : [0, 1] \rightarrow \mathcal{C}$ can be represented as a sequence of continuous configurations connected from q_{init} to q_{goal}, where q_{init} is the starting configuration and q_{goal} is the target configuration. a path is considered valid if it satisfies $\sigma(t) \in \mathcal{C}_{free}$ $\forall t \in [0, 1]$, where $\sigma(0) = q_{init}$, $\sigma(1) = q_{goal}$.

Problem 1. Given the robot configuration space $\mathcal{C}$, the initial configuration q_{init}, and the target configuration q_{goal}, a motion planning problem can be expressed as a triplet $(\mathcal{C}, q_{init}, q_{goal})$. The goal of the path planning algorithm is to construct a feasible path that satisfies Definition 1 so that the robot can move from q_{init} to q_{goal} without collision.

Problem 2. In semi-structured environments, the robot usually needs to do some repetitive manipulation tasks (e.g., moving the target objects from a starting location to a target location in the workspace $\mathcal{W}$); we define such tasks as $T = \{\tau_1, \tau_2, \cdots, \tau_n\}$. Repetitive path planning problem is to solve the planning problem $(\mathcal{C}, f(x_{init,i}), f(x_{goal,i}))$ for all $\tau_i \in T$, where x_{init} and x_{goal} are the initial and goal pose of the manipulator end effector in the workspace. $f(\cdot)$ is the function that map a pose $x \in \mathcal{W}$ of the manipulator end-effector to a possible configuration $q \in \mathcal{C}_{free}$, which can be achieved by inverse kinematics.

3 Proposed Method

3.1 Approach Overview

There is a distinguishing feature for manipulation tasks in semi-structured environments: the planning problem for each task instance is similar, and the planned paths may lie in some specific regions of the configuration space. It is preferable to utilize the information contained in previous planning tasks to accelerate subsequent planning problems. In this work, an experience-based motion planning method is proposed to solve the repetitive path planning problem in semi-structured environments. An overview of the proposed method is shown in Fig. 2. First, we collect some planned paths for similar tasks in advance to build a path library. Then the Gaussian mixture model (GMM) is used to extract the path features from the path library. After obtaining the GMM, an adaptive sampler combining a uniform sampling strategy and a GMM-based sampling strategy is constructed. This can guide the planner to explore towards task-relevant regions where the expected path is likely to exist. In addition, the exploration of the configuration space by similar tasks in a semi-structured environment contains valuable information. In this paper, we adopt an experience graph to capture the exploration of configuration space by all previous tasks, and subsequent planning tasks can retrieve locally valuable information from it to avoid repeated exploration.

For a given planning request, initialize the initial and goal trees according to the given initial and goal configurations. The adaptive sampler is then employed to guide the tree to explore the configuration space while attempting to exploit valuable information from the experience map. After obtaining the initial solution, a path pruning algorithm is used to remove redundant waypoints to obtain a shorter path. After the current task is completed, update the experience graph

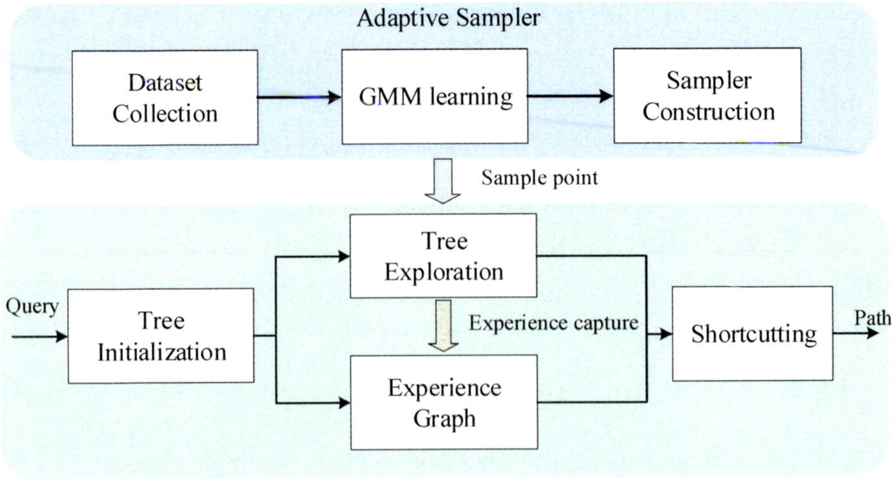

Fig. 2. Approach overview.

according to the new exploration of the configuration space by the current task. The remainder of this section will introduce the adaptive sampler, the experience graph, and the implementation details of the proposed method.

3.2 Adaptive Sampler

Let $\mathcal{A} = \{\sigma_1, ..., \sigma_L\}$ be the path library, which refers to the set of demonstration paths collected from previous similar planning tasks. Each $\sigma_i \in \mathcal{A}$ is a set of configuration points, so the path library can also be represented as $\mathcal{A} = \{q_1, ..., q_N\}$, which represents the set of configuration points for all collection paths. Let Ω be the task-relevant regions where the path may exist. The learning phase of GMM is shown in Fig. 3, and the Gaussian mixture model used to learn these task-relevant regions from the path library can be represented as:

$$P(q|\theta) = \sum_{k=1}^{K} \pi_k \phi(q|\theta_k) \tag{1}$$

where π_k stands for the weight of the k-th Gaussian component and satisfies $\pi_1 + \pi_2 + ... + \pi_K = 1$, $\pi_k \geq 0$. $\phi(q|\theta_k)$ represents the probability density function of the k-th Gaussian component, which can be denoted as:

$$\phi(q|\theta_k) = \frac{1}{(2\pi)^{\frac{d}{2}} |\Sigma_k|^{\frac{1}{2}}} exp^{-\frac{1}{2}(q-\mu_k)^T \Sigma_k^{-1}(q-\mu_k)} \tag{2}$$

where $\theta_k = (\mu_k, \Sigma_k)$, μ_k and Σ_k are mean value and covariance matrix of the k-th Gaussian component, respectively, d is the dimension of the configuration space. The parameters of the GMM are obtained by the EM algorithm, and the solution steps are as follows: First, The Akaike Information Criterion(AIC) is used to determine the optimal number of components of GMM; secondly, the K-means clustering algorithm is used to obtain the initialization parameters of the GMM; then E-step and M-step iterate repeatedly until the required GMM parameters are obtained.

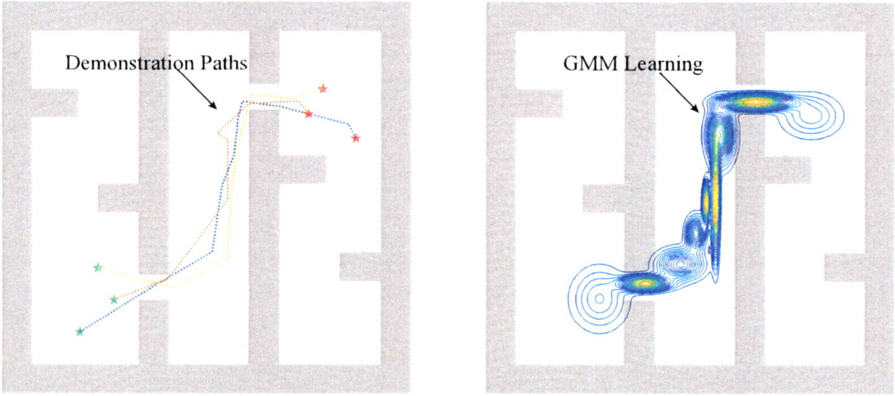

Fig. 3. GMM Learning from demonstration path.

After using the GMM to obtain the probability distribution of the teaching data, an adaptive sampler is constructed that combines a uniform and a GMM-based sampling strategy. It can make the sampling points fall into the task-related region with a certain probability. The pseudo-code of the adaptive sampler is shown in Algorithm 1, where k represents the proportion of the GMM-based sampling strategy in the entire sampling process.

Algorithm 1: Adaptive_Sampler()

```
rand ← Randnum(0, 1)
if rand ≤ k then
  | q_rand ← UniformSampler()
else
  | q_rand ← GMMSampler()
end
return q_rand
```

3.3 Experience Graph

In semi-structured environments, manipulation tasks are usually highly repetitive. These similar tasks will lead to similar planning problems, and previous tasks' exploration of configuration spaces contains valuable information. Subsequent planning tasks can exploit this information to avoid exploring the configuration space repeatedly. In this work, we utilize an experience graph to capture the exploration of configuration spaces by previous tasks, as shown in Fig 4. The gray blocks, red pentagram, and blue pentagram represent obstacles, initial point, and goal point, respectively. The goal of motion planning is to find a collision-free path for a robot from the initial point to the goal point. The solution process of the proposed method is as follows: First, Given a path planning problem, initialize the initial and goal tree with the given initial and goal point. When the planner runs for the first time, the experience map is empty; two trees explore the configuration space simultaneously with the adaptive sampler. When the two trees are connected, a feasible path can be returned, and the exploration of configuration space by two trees will be kept in the experience graph without collision detection (see Fig. 4(1)). For subsequent tasks, the trees only need to explore previously unexplored regions and use the information stored in the experience graph. If both trees are close to the experience graph, the shortest path in the experience graph can be quickly returned with the A* search algorithm and lazy collision checking strategy. Moreover, even if the appearance of new obstacles (the black block in Fig. 4(3)) in the environment makes the previously calculated path invalid, a new candidate path can be obtained quickly due to the introduction of a lazy collision checking scheme. Finally, the new exploration of the configuration space by the current task will be merged into the experience graph to capture more information (see Fig. 4(4)).

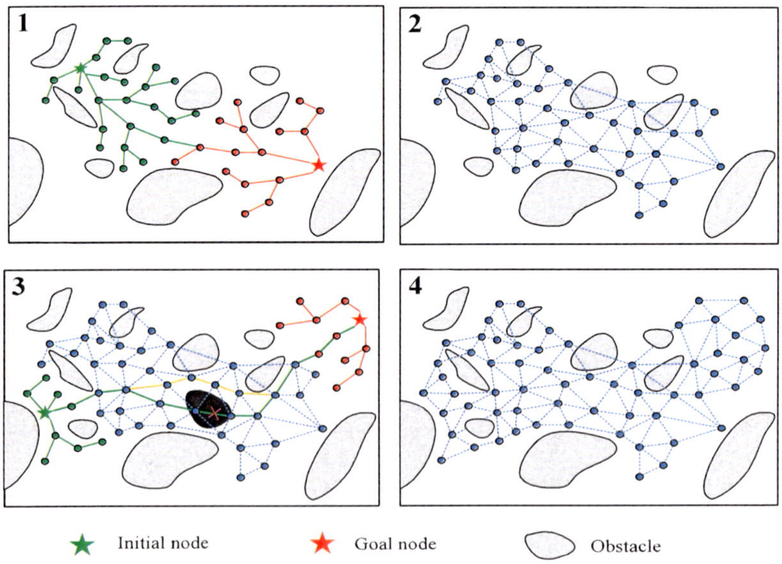

Fig. 4. An Illustration of experience capture and retrieval.

3.4 Implementation

In Sects. 3.2 and 3.3, we introduce two crucial experience capture modules of the proposed method. In this subsection, We will introduce the implementation details of the proposed method, and The entire procedure is shown in Algorithm 2. First, the experience graph is initialized to be empty when the motion planner runs for the first time (Line 30). For each planning task, the path is obtained by calling the function *EBRRT**$(\cdot)$ (Line 32). This function contains three input parameters: G is the experience graph that contains valuable information about configuration spaces, and $x_{start,i}$ and $x_{goal,i}$ are the robotic arm's starting pose and goal pose for the ith task, respectively. $f_q(\cdot)$ is the function that maps the configuration of the end-effector in the workspace $\mathcal{W}$ to a possible robot joint configuration in *C-space*, which can be achieved by inverse kinematics method.

The details of the process *EBRRT**$(\cdot)$ is shown in Line 1–Line 29. It will run iteratively until an initial solution is found within the time budget δ_t (Line 4–Line 28). After obtaining the initial solution, the path pruning algorithm eliminates redundant points to obtain a shorter path (Line 33). The subroutine EBRT can be described as follows: First, initialize the start tree $\mathcal{T}_a$ and the goal tree $\mathcal{T}_b$ with the input start and target point (q_{start}, q_{goal}) as the root node, respectively, the extension flags *extend* () of the two trees are set to true and take $\mathcal{T}_a$ as the current extension tree. The extension flag represents whether the current extension tree should continue to explore in the configuration space; if false, stop its exploration process. If the extension flag of the current tree is true, the adaptive sampler is used to generate a random configuration q_{rand} in the configuration space *C-space*, the current tree will attempt to extend towards q_{rand} and generate a new configuration point q_{new} (this extension process is described in detail in [14]). If $q_{new} \neq \emptyset$, the current tree will try to connect with another tree. If the two trees are successfully connected, a path σ from the starting point to the goal point is found, and the whole iterative process is stopped. Otherwise, the current tree will try to connect with the experience graph to retrieve locally valuable information to avoid exploring the configuration space repeatedly. If the current tree can be connected to the experience graph G, stop the expansion of the current tree and set the explore flag *extend* to False (Lines 8–15).

When both trees are connected to the graph G, a path search method is performed to find a locally feasible path in the graph G between two connected vertices. In this paper, the graph search method is the lazy collision checking strategy and the A* graph search method. The validation of the edges in the graph is detected lazily rather than immediately, which can save unnecessary computational time and endows the graph G to adapt to changes in the environment. If such a local exit, the final path solution that connects q_{start} and q_{goal} through $\mathcal{T}_a$, G and $\mathcal{T}_b$ can be obtained. Otherwise, set $\mathcal{T}_a.extend$ and $\mathcal{T}_b.extend$ to be True, indicating that the two trees' growth has been restored (Lines 18–24). After each iteration, swap $\mathcal{T}_a$ and $\mathcal{T}_b$ so that the two trees alternately explore the configuration space. Furthermore, after the path of the current task is found, the exploration of the configuration space by $\mathcal{T}_a$ and $\mathcal{T}_b$ will be incrementally merged into G to enrich the experience graph.

Algorithm 2: Experience-guide BiRRT*

```
Function EBRRT*(G, q_start, q_goal):
    // initialise T_a, T_b
    T_a.V ← {q_start}, T_a.E ← ∅, T_a.extend = True
    T_b.V ← {q_goal}, T_b.E ← ∅, T_b.extend = True
    while within time budget δ_t do
        if T_a.extend = True then
            q_rand = Adaptive_Sampler()
            q_new ← TreeExtend(T_a, q_rand)
            if q_new ≠ ∅ then
                if ConnectTree(T_b, q_new) then
                    σ ← find a path from T_a and T_b
                    break
                end
                if ConnectGraph(G, q_new, T_a) then
                    T_a.extend = false
                end
            end
        end
        if T_a.extend = T_b.extend = False then
            // T_a and T_b are connected to G
            σ ← LazyGraphSearch()
            if IsFeasible(σ) then
                break
            else
                // restart the exploration of two trees
                T_a.extend = T_b.extend = True
            end
        end
        swap(T_a, T_b)
    end
    G ← UpdateGraph(G, T_a, T_b)
    return σ
G = (V ← ∅, E ← ∅)
for τ_i ∈ T do
    σ_i ← EBRT*(G, f_q(x_start,i), f_q(x_goal,i))
    PathPruning(σ_i)
end
```

4 Experiments

Two simulation experiment scenarios are constructed to verify the proposed method's effectiveness. One is the path planning problem of a particle robot in a 2D grid map, and the other is a 7-DOF manipulator scenario in which the robot performs repetitive pick-and-place tasks. All experiments are tested on an Intel i5 Linux machine with 4 × 2.5 GHz cores and 16 GB of RAM. The rest of this section describes the setup and results of the experiments.

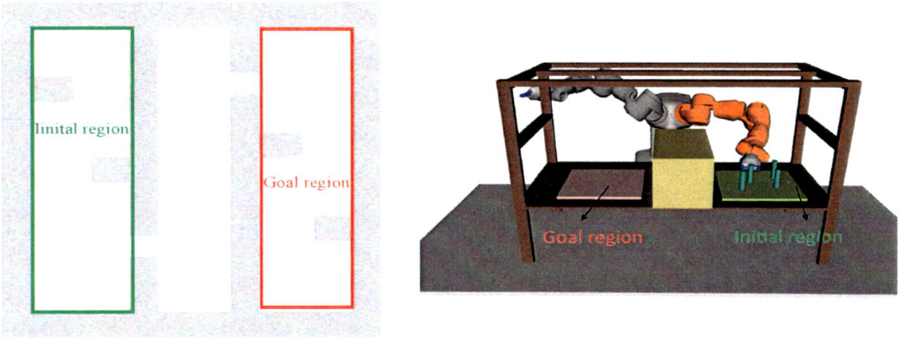

(a) (b)

Fig. 5. Experimental scenarios: a 2D simulation scenario with narrow passage (a), a 7-DOF manipulator simulation scenario where the robot need to perform repetitive pick-and-place tasks (b).

4.1 2D Simulation

A 2D simulation environment of size 500×500 is constructed to verify the proposed method's effectiveness on low-dimensional space planning problems, as shown in Fig. 5(a). Grey blocks represent obstacles in the environment, and white areas represent free configuration spaces. The experiment consisted of 6 similar planning tasks that the robot needs to move from the starting point to the goal point, where the starting point and the target point are randomly generated in the starting and target areas (green region and red region in Fig. 5(a), respectively. We compare the proposed method with some classic and representative methods (e.g., RRT [11], BITTR [13], RRT* [12], BiRRT* [14]); each planning task is solved with five different planners and repeated 10 times. All the experiments are implemented in MATLAB2020b.

4.2 7-DOF Manipulator

This paper built a 7-DOF manipulator platform for repetitive pick-and-place operation tasks to test the effectiveness of the proposed method for high-dimensional planning problems, as shown in Fig. 5(b). The robot needs to pick the blue object from the starting area (green region) and place it in the target area (pink region) while avoiding obstacles. Most objects in the environment are relatively static, and only slight environmental changes occur due to the manipulator grasping and moving objects. All planning tasks are performed in joint space (C-space), and the inverse kinematics is used to map the initial and goal pose of the end-effectors in the workspace $\mathcal{W}$ to the corresponding start and target configuration in the configuration space. All experiments are performed on the ROS operating system with the MoveIt! framework; the collision checking library is FCL. This experiment consists of a total of 4 picking and 4 placing tasks. Each task is repeated 10 times with five different planners, totaling 400 trajectories.

4.3 Results

Figure 6 shows the planning process for two consecutive tasks in a 2D environment. For the first planning task, the proposed method behaves similarly to BiRRT* [14] but adopts an adaptive sampling strategy, which makes the sampling points more likely to fall into the task-related region (see in Fig. 6(a)). At the same time, the tree's exploration of the configuration space is preserved in an experience graph (see Fig. 6(b)). For subsequent planning tasks, the tree only needs to explore regions that have not been explored before. The exploration of configuration space by previous tasks can be reused to help the current planning task quickly find a path (see Fig. 6(c)). Even if the environment changes (suppose this black object is a new obstacle that occurs in the environment), a feasible candidate path can be found quickly due to the lazy collision detection strategy (see Fig. 6(d)).

We evaluate the performance of different algorithms in terms of path cost and planning time metrics. The planning time represents the time required for the planner to find a feasible path from the starting point to the goal point, and the path cost reflects the total path length from the starting point to the goal point. Figure 7 depicts the results of various planners as time and cost boxplots across the evolution of tasks 1–6 in the 2D simulation scenario. When performing task 1, the proposed method is similar to BiRRT* [14] but takes less time and is more stable since the adaptive sampler can guide the tree to explore task-relevant regions, avoiding blind sampling in the entire configuration space. At the same time, the exploration of configuration space by the trees is summarized in an experience graph. It can obtain solutions faster than other planners for subsequent tasks by exploiting experience gleaned from previous similar tasks. RRT [11] and BiRRT [13] can also rapidly get solutions but do not guarantee path quality and are unstable. On the contrary, the paths planned by EBRRT* are always the shortest among all methods, as shown in the cost boxplots (see Fig. 7(b)). This is due to the fact that the proposed method inherits the asymptotically optimal property of BiRRT* [14], and the redundant points in the path are removed with a path pruning procedure.

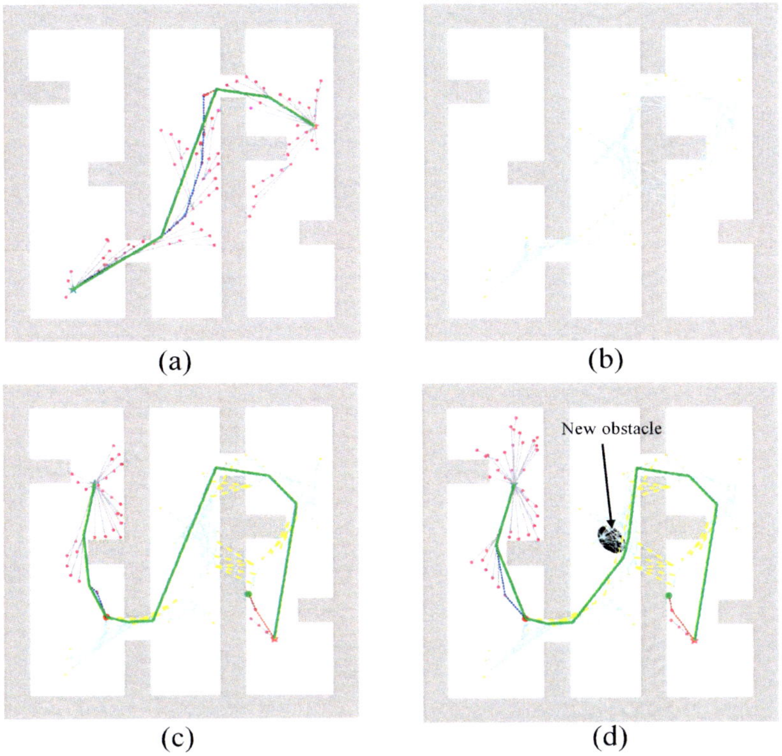

Fig. 6. Two consecutive planning processes in a 2D simulation scene.

Table 1. Mean results of different planners on the 1*th* and 2*th* pick-and-place tasks in 7-DOF manipulator scenario

	#TASK	RRT	BiRRT	RRT*	BiRRT*	Ours
Time (s)	$Pick^{1th}$	2.069	**0.994**	9.699	7.048	5.118
	$Place^{1th}$	3.580	1.082	14.218	8.798	**0.546**
	$Pick^{2th}$	2.160	0.928	9.098	6.833	**0.425**
	$Place^{2th}$	2.142	1.054	8.724	5.162	**0.774**
Cost	$Pick^{1th}$	19.017	20.336	11.727	12.471	**11.125**
	$Place^{1th}$	18.209	19.054	13.747	13.329	**11.998**
	$Pick^{2th}$	17.286	20.702	12.320	13.922	**10.419**
	$Place^{2th}$	17.343	20.701	12.061	12.047	**10.468**

Table 1 displays the experimental results of the 7-DOF manipulator scenario in terms of planning time and path cost. The first and second pick-and-place tasks are included to illustrate the performance of different planners for solving repetitive path planning problems in semi-structured environments. The EBRRT* has a higher time cost than RRT and BiRRT for the first pick

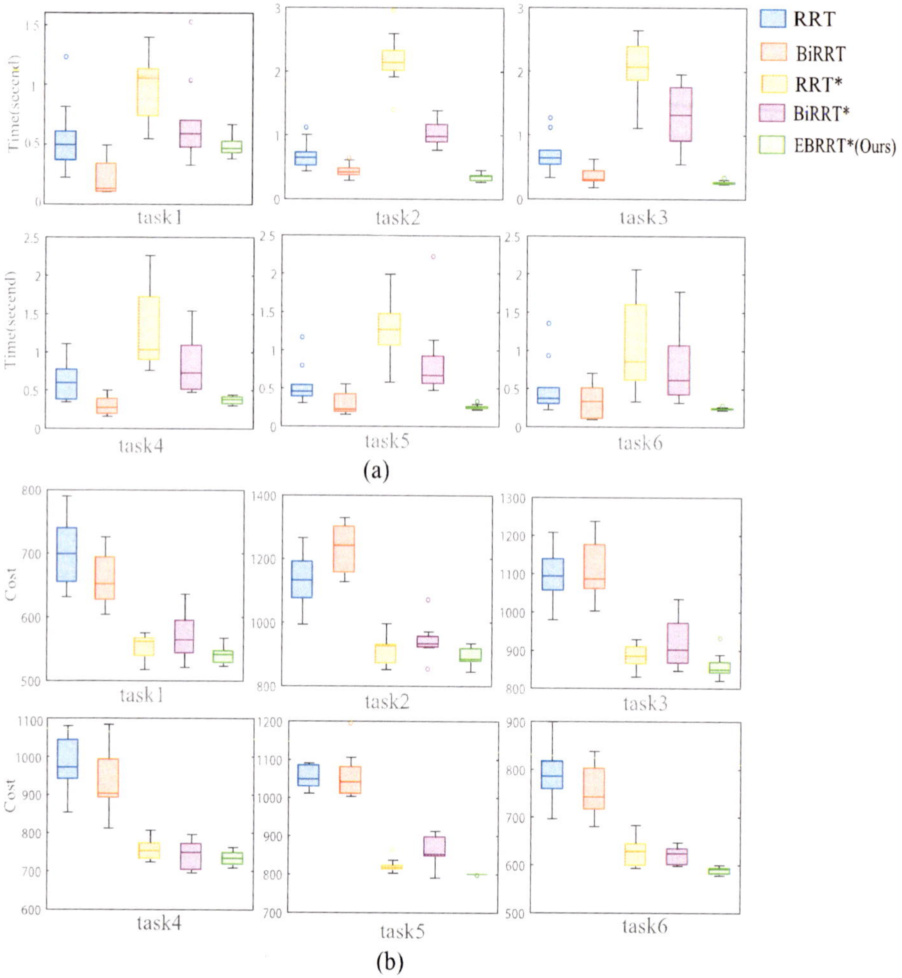

Fig. 7. Planning time and path length for 2D simulation environment.Each task instance is planned using five different planners, repeated 10 times.

task. However, in subsequent tasks, EBRRT* can obtain solutions more quickly. Because the underlying configuration spaces of these tasks are similar, and the exploration of the configuration space by previous tasks is saved in the experience graph, subsequent tasks can reuse the valuable information stored in the experience graph to speed up problem-solving (as shown in Fig. 8). In addition, it can be seen from the path cost metric that the path planned by EBRRT* is always the smallest among these planners. Although the asymptotically optimal motion planner (RRT*, BIRRT*) can also obtain a relatively short path but takes too long. In summary, these tree-based motion planners plan each task from scratch without considering knowledge transfer among similar tasks. On the contrary, the proposed method can explore the configuration space as fast

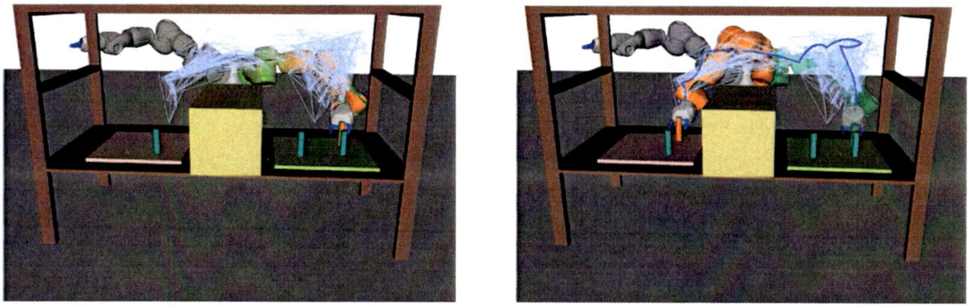

Fig. 8. Two consecutive pick-and-place tasks in a 7-DOF manipulator scenario.

as the tree-based planner while exploiting the experience captured from previous similar tasks to speed up the solution of the path-planning problem. The experimental results show that the proposed method performs well in solving the repetitive path planning problem.

5 Conclusion

This paper presented an experience-guide path planning method (EBRRT*) combining a GMM-based adaptive sampler and an experience graph, which can automatically capture valuable information from past motion plans. For subsequent subtasks, it can guide sampling in task-related regions and retrieve locally valuable information from the experience graph via lazy collision detection strategy and A* graph search algorithm to quickly find a high-quality shorter path. Furthermore, the proposed method is visualized and verified in a 2D simulation environment and a 7-DOF manipulator scenario. The results of both experiments show that the proposed method can find a high-quality path faster than other advanced tree-based planners for repetitive path planning in semi-structured environments. Future work would like to combine the proposed method with task planning to solve more complex robot planning problems.

References

1. Islam, F., Salzman, O., Agarwal, A., et al.: Provably constant-time planning and replanning for real-time grasping objects off a conveyor belt. Int. J. Robot. Res. **40**(12–14), 1370–1384 (2021)
2. Ying, K.C., Pourhejazy, P., Cheng, C.Y., et al.: Deep learning-based optimization for motion planning of dual-arm assembly robots. Comput. Indus. Eng. **160**, 107603 (2021)
3. Wang, H., Zhao, Q., Li, H., et al.: Polynomial-based smooth trajectory planning for fruit-picking robot manipulator. Inform. Process. Agricult. **9**(1), 112–122 (2021)
4. Wang, Y.: Motion planning of space robot. In: Space Robotics. SST, pp. 35–60. Springer, Singapore (2021). https://doi.org/10.1007/978-981-15-4902-1_3
5. Dai, Y., Xiang, C., Zhang, Y., et al.: A Review of spatial robotic arm trajectory planning. Aerospace **9**(7), 361 (2022)

6. Chen, N., Zhang, Y., Cheng, W.: Space detumbling robot arm deployment path planning based on Bi-FMT* algorithm. Micromachines **12**(10), 1231 (2021)
7. Islam F, Paxton C, Eppner C, et al.: Alternative paths planner (app) for provably fixed-time manipulation planning in semi-structured environments. In: Proceedings of IEEE International Conference on Robotics and Automation, pp. 6534–6540. IEEE, Xi'an (2021)
8. Pairet, É., Chamzas, C., Petillot, Y., et al.: Path planning for manipulation using experience-driven random trees. IEEE Robot. Autom. Lett. **6**(2), 3295–3302 (2021)
9. Kavraki, L.E., Kolountzakis, M.N., Latombe, J.C.: Analysis of probabilistic roadmaps for path planning. IEEE Trans. Robot. Autom. **14**(1), 166–171 (1998)
10. Bohlin, R.; Kavraki, L.E.: Path planning using lazy PRM. In: Proceedings of the IEEE International Conference on Robotics and Automation, pp. 521–528. IEEE, San Francisco (2000)
11. LaValle, S.M.: Rapidly-exploring random trees: a new tool for path planning. Technical Report, Computer Science Department, Iowa State University, Ames, IA, USA (1998)
12. Karaman, S., Frazzoli, E.: Sampling-based algorithms for optimal motion planning. Int. J. Robot. Res. **30**(7), 846–894 (2011)
13. Kuffner, J.J., LaValle, S.M.: RRT-connect: an efficient approach to single-query path planning. In: Proceedings of IEEE International Conference on Robotics and Automation, pp. 995–1001. IEEE, San Francisco (2000)
14. Klemm, S., Oberländer, J., Hermann, A., et al.: Rrt-connect: faster, asymptotically optimal motion planning. In: Proceedings of 2015 IEEE International Conference on Robotics and Biomimetics (ROBIO), pp. 1670–1677. IEEE, Zhuhai (2015)
15. Szkandera, J., Kolingerová, I.: Iterative solution for the narrow passage problem in motion planning. In: Proceedings of International Conference on Computational Science, pp. 219–232. Springer, Cham (2022)
16. Wang, X., Ma, X., Li, X., et al.: Target-biased informed trees: sampling-based method for optimal motion planning in complex environments. J. Comput. Des. Eng. **9**(2), 755–771 (2022)
17. Cai, P., Yue, X., Zhang, H.: ADD-RRV for motion planning in complex environments. Robotica **40**(1), 136–153 (2022)
18. Gammell, J.D., Barfoot, T.D., Srinivasa, S.S.: Batch Informed Trees (BIT*): informed asymptotically optimal anytime search. Int. J. Robot. Res. **39**(5), 543–567 (2020)
19. Ichter, B., Harrison, J., Pavone, M.: Learning sampling distributions for robot motion planning. In: Proceedings of IEEE International Conference on Robotics and Automation, pp. 7087–7094. IEEE, Brisbane (2018)
20. Wang, J., Chi, W., Li, C., et al.: Neural RRT*: learning-based optimal path planning. IEEE Trans. Autom. Sci. Eng. **17**(4), 1748–1758 (2020)
21. Shah, N., Srinet, A., Srivastava, S.: Learning and using abstractions for robot planning. arXiv preprint arXiv:2012.00658 (2020)
22. Cheng, Q., Zhang, W., Liu, H., et al.: Research on the path planning algorithm of a manipulator based on GMM/GMR-MPRM. Appl. Sci. **11**(16), 7599 (2021)
23. Qiu, Q., Cao, Q.: Motion planning in semistructured environments with teaching roadmaps. Intell. Serv. Robot. **13**(3), 331–342 (2020). https://doi.org/10.1007/s11370-020-00316-9

Author Index

© The Editor(s) (if applicable) and The Author(s), under exclusive license to Springer Nature Switzerland AG 2023
M. Ceccarelli et al. (Eds.): TORVEASTRO 2023, MMS 130, p. 193, 2023.
https://doi.org/10.1007/978-3-031-28447-2